Ajay &

Great meeting
you @ Prenotice

E a

Social Engagement for Customer Care

a practitioner's guide to driving loyalty on the social web

Covering Social Engagement Strategy, Essential Discovery Steps, Etiquette for Outreach, Profiling Strategies, Agent Assistance, Multi-Tiered Clustering Strategies, and Best Practices for the Social Command Center.

By Edwin Margulies
Edited by Lance Fried
Foreword by Chad McDaniel, President and Founder of Execs In The Know

Social Engagement for Customer Care

A practitioner's guide to driving loyalty on the social web

Published by Sterling Press, Dallas, TX United States

Copyright © 2013 by Edwin Margulies

ISBN 978-0-9749935-8-4
Manufactured in the United States of America
Printed in USA by Brandt Doubleday Inc.,
Kalamazoo, Michigan
First Edition, May 2013

ACKNOWLEDGEMENTS

There are a lot of people to thank for making this book possible. I'd like to thank my colleagues at SoCoCare. Let me start by thanking Lance Fried for his support as we prepared this manuscript. Lance was especially helpful in providing the guidance necessary as my editor. His input and cajoling on the draft through final copy was essential. Special thanks also go to Sarah Rolfing, our marketing wizard, who is responsible for the cover art, diagrams and for producing our blog series. Sarah's support on my blog has been invaluable and helped to inspire this book.

Thanks also to our visionary development staff and their leader and CTO Ran Ezerzer, whom I have worked with for many years in bringing great products to market. I would also like to thank our director of implementation and program management, Michael Haisten for his great contributions in the area of discovery and best practices.

Edwin Margulies
Van Alstyne, Texas, May 2013
ed@sococare.com

Table of Contents

FOREWORD

Who Should Read This Book and Why

Consumer behavior and engagement with your brand has changed exponentially within the last five to seven years alone. Think how Social Networking has changed the numerous ways we personally engage, network and share communication within our own social networks. We all now have the unique ability to "self-publish" and get our thoughts, ideas and feelings on the worldwide web in seconds.

The public, convenient and speedy nature of our daily interactions on social media doesn't stop at brands. In fact social customer service offers consumers benefits they haven't been able to get from the companies they buy from since the days of the local corner store. Today, in the social age, we're living in a global community where everyone has a voice. We should be listening carefully.

Consumers are bringing more and more queries, questions and complaints to social channels. It is clear that businesses ignoring customers through social channels will continue to become increasingly rare.

A customer service strategy on social channels needs to be formed and implemented urgently if businesses do not want to be left in the wake of their competitors.

More than ever, consumers have a global voice. Comments and sentiment on your brand can spread infinitely more quickly than in the days of word-of-mouth. More than ever, sentiment can quickly spread and reverberate through the web, causing real damage to brands.

A number of businesses struggle to understand how to effectively create a social channel strategy that can prove ROI. Equally, many overlook the importance of engaging properly.

As the shift in consumer behavior continues to evolve, brands should not be fearful of the customer. Yes, the customer has a new and often very loud voice; however, brands still have the unique ability to control experience - if they are willing to accept this "new" norm. Take control by ensuring you create social channel policies, guidelines and strategies for communicating with customers to engage with them rather than ignore them.

Social customer engagement can appear daunting to many because of its inherently unruly nature, with all communications laid bare for the world to see.

If you're reading this book, you have probably already decided that your company wants to be part of the solution and has chosen to listen, engage and help its customers via social channels.

You should certainly read this book if you don't know where to start in Social Engagement for Customer Care. Its pages are crammed with essential discovery tips, best practices, and methods for getting your social command center up and running.

Even if you are a veteran of social care, this book will help you fine-tune your practices and maybe you will learn some new ideas for customer outreach on social channels.

It is simply not good enough to set up an online presence and think that it can be used as a one-way medium, sending out 'special offer' messages to legions of loyal customers.

Social channel engagement is all about taking part in conversations with people and communicating your strong desire to connect with your customers. Leaving customer concerns unaddressed could prompt anger and you could risk severing ties with people who might have a long-standing relationship with your brand.

You should also read this book if you are an executive involved in customer care, brand management or voice of the customer initiatives.

This non-technical book will help you:

- Establish a strategy for engaging customers over social channels

- Increase the productivity of your social care team at your company

- Provide superior customer service to clients using social channels to communicate with your company

- Give your customers many more reasons for dealing with your company

- Save your company thousands of dollars on customer-facing initiatives with proven best practices for social engagement

If you are new to Social Engagement for Customer Care, welcome to the brave new world of the "social contact center." I hope this book will help you to improve your engagement skills and practices so you can build and maintain a loyal following on the social web.

May 2013

Chad McDaniel
President and Founder
Execs In The Know –
Advocates of the Customer Experience Executive
www.execsintheknow.com
chad@execsintheknow.com

1. Strategies for Social Customer Care

Whether your are just getting started with your social engagement for customer care initiative, or trying to make improvements to an existing effort, it's a good idea to pause and take time to cement your strategy.

A useful methodology for formulating your approach is to list your most important observations and conclusions about your customers' activity on the social web.

Looking at social post items that mention your brand or a competitor's brand can help you in making these observations.

If you already have native Facebook or Twitter accounts, making observations is as easy. All you have to do is "listen" to what people are saying about your brand. You can also do searches on key words to establish conversation topics amongst your customers.

There are a variety of tools you can use to observe social mentions of your brand. These are simple "listening" tools such as HootSuite or TweetDeck, for example. You can also use more advanced tools such as full-blown engagement platforms such as SoCoCare's Social CIM. You can do cursory observations before these systems are fully provisioned. Either way, before you jump into a full-blown deployment, I recommend you take the time to make some essential observations first.

Essential Observations

In making observations, first pick a specific, recent time frame to study. For example, if you have a seasonal business, you might want to take a look at time frames in which the bulk of transactions happen. Let's say you do

2

70% of your business between mid-November and mid-January. Focus on that time frame to garner the largest amount of observations on your brand.

Second, attempt to categorize themes you can easily recognize. These themes often fall into obvious buckets such as "sales," "service," "complaints," or "endorsements." Third, tally the number of social post items that occur for each of them inside of that time frame.

If you are running a grocery store chain, you might observe themes having to do with customer service, quality of produce, or sale items for example. You can keep track of your observations using a simple table or spreadsheet as pictured below:

Table: Social Theme Tally

Timeframe	Count	General Content Theme
Nov. 15 – Jan 15	2,893	Non-news brand mentions
Nov. 15 – Jan 15	1,257	Brand endorsements or referrals
Nov. 15 – Jan 15	8,544	News items with brand mentions
Nov. 15 – Jan 15	945	Complaints (customer service)
Nov. 15 – Jan 15	1,134	Complaints (product quality)
Nov. 15 – Jan 15	900	Questions @ store hours, sales.
Nov. 15 – Jan 15	543	Non-specific rants

Clearly, a tally of observations such as this could represent a significant amount of work. And it can be tedious. But I encourage you to put the work into detailing your observations because this step is essential in setting up your strategy. You can use native interfaces to do some of this work and you can also use more advanced analytics tools to achieve the same result.

Another approach if you have a lot of volume is to do random sampling to establish a percentage of each theme and then estimate the number of similar themes against the total number of posts during that time frame. Although not exact, this will yield decent results so long as your sample size is large enough — say maybe 500 social post items.

One thing to keep in mind during this exercise is to make observations on a variety of social web sources. For example, if all you look at is tweets, and you don't look at Facebook or blogs — you may get a skewed result. That is to say that some source feeds may run "hotter" than anothers sentiment-wise — especially if there are a lot of re-tweets or re-posts of a specific item.

Some practitioners find that entire sources are off the mark from a customer care perspective.

Take, for example news articles. These often contain rich information on a product or brand, but often do not contain anything actionable for a customer care agent to do anything about. One way to memorialize these observations is to plot them on a simple chart using one axis for volume of post items and the other axis representing your view of relevancy in the customer care area.

You can create "bubbles" for each theme and group the bubbles based on their relative volume and relevancy. Then you can draw a dotted box around the themes most likely to yield actionable content for your customer care team. Such a chart would look like this based on the example observations:

Figure: Chart based on Volume and Relevancy

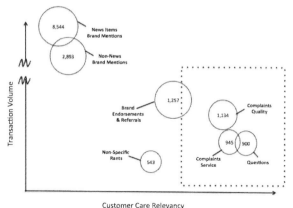

5

Now it's time to apply some interim conclusions to the tally you have created. For example, you may conclude that there are a lot of non-news brand mentions, but the bulk of them are not actionable from a customer care standpoint. There may be some good data in that category in terms of mining for marketing information or trends, but nothing your customer care team can help with. Likewise, you may conclude that although you are thrilled with brand endorsements or referrals, your care team will not be geared-up to help those folks.

(On the other hand your sales team may want to pick through the referrals and kudos to provide loyalty points or other incentives to social authors who are supporting your brand).

These simple conclusions will help you with a process of elimination. You can cross-off themes from the list so it is easier to further characterize and establish strategies on what's left.

Table: Themes & Process of Elimination

Timeframe	Count	General Content Theme
~~Nov. 15 – Jan 15~~	~~2,893~~	~~Non-news brand mentions~~
~~Nov. 15 – Jan 15~~	~~1,257~~	~~Brand endorsements or referrals~~
~~Nov. 15 – Jan 15~~	~~8,544~~	~~News items with brand mentions~~
Nov. 15 – Jan 15	945	Complaints (customer service)
Nov. 15 – Jan 15	1,134	Complaints (product quality)
Nov. 15 – Jan 15	900	Questions @ store hours, sales.
~~Nov. 15 – Jan 15~~	~~543~~	~~Non-specific rants~~

Table: Last Themes Standing

Timeframe	Count	General Content Theme
Nov. 15 – Jan 15	945	Complaints (customer service)
Nov. 15 – Jan 15	1,134	Complaints (product quality)
Nov. 15 – Jan 15	900	Questions @ store hours, sales.

It's also a good idea to revisit the "last themes standing" data from time to time. I also suggest making periodic observations that could lead to completely different conclusions.

The social web is in constant change, so not every topic will be persistent. I therefore recommend repeating these observations steps at least once every three months.

In our sample process so far, we've concluded that the content themes for customer care deal with complaints about customer service or quality. The same customer care concept involves questions about store hours, sales and other service-related issues. Now let's take a look at how these themes can be further characterized.

You can now take a look at the general sentiment of these posts and what percentages of them are not actionable or "spam." In the context of a customer service scenario, any social post item that is not a candidate for outreach is "spam." That is to say that any post item that does not provide an actionable path for the social care agent to do outreach or help in some way is "spam."

It is important to understand that just because a care specialist would characterize a post item as being "spam," that does not mean that the enterprise as a whole cannot benefit from paying attention to the post.

Even non-specific rants about a brand can be of some use to a brand manager to track sentiment trends, even if they are not actionable from a customer care perspective. Ditto endorsements or referrals that beg for sales follow-up in the area of cross-sell or up-sell.

Table: Sentiment and spam View of Themes

General Content Theme	% Angry	% Neutral	% Happy	% spam
Complaints (customer service)	70	20	10	10
Complaints (product quality)	68	23	9	10
Questions	9	90	1	5

Since the focus of this book is social engagement for customer care (with an emphasis on customer care) we will not be addressing the broader disciplines of marketing, campaigning, and brand analysis.

Setting Interim Goals

Before we jump into strategy it is first important to address the basic goals you want to achieve.

Now that you have established themes based your observations, you can match those themes up to the chief disciplines that make up customer care.

These include: a) Issue Resolution (helping to resolve problems for customers); b) Product Feedback (requests for features, capabilities, new offerings, etc.); c)

Knowledge Transfer (answering general, product-specific or technical questions); d) Account Servicing (billing, renewals, membership); and e) Customer Retention (complaints, escalations, etc.).

Clearly, the chief disciplines for a customer care operation are going to change depending on the industry you are in. These customer care disciplines will also change from company to company.

You can jot down the ones that make sense for you and then link them to the themes you have uncovered in the earlier exercise. One way to plot these care disciplines graphically versus your observed themes is like this:

Figure: Care Discipline Mapping to Themes

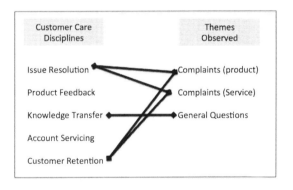

Formulating a Strategy

Now you are ready to begin formulating a basic strategy for your customer care initiative. Here are some ideas to get on the table in formulating your strategy.

Use 3x5 cards or a whiteboard so you can arrange the words from the figure below. This lets everyone on your team participate in the strategy process. Each row covers a specific aspect of how you could run your social care initiative.

Your corporate culture, business rules, and human resource norms should be taken into consideration. Ditto the profile of your customers.

Table: Social Care Strategy "Bingo Card"

Skill Focus:	Generalist	OR	Hybrid Teams	OR	Specialist
Approach:	Quality Focus	OR	Business Rules	OR	Volume Focus
Content:	Pure Social	OR	KB Referral	OR	Channel Conversion
Key Drivers:	Sentiment	OR	Influence	OR	Loyalty
Supervision:	Tight Control	OR	Quarterbacking	OR	Agent Autonomy
Metrics:	Group SLA	OR	Agent KPI	OR	Disposition Trends

Let's take a look at each of the aspects that can go in to your specific strategy. This will help to get the dialog going with your team.

Skill Focus

There are many factors that go in to the strategy for skill focus. One factor is the number of people you have on your team. Another factor is the background and domain expertise of the individuals on the team.

The smaller the team, the more there is a tendency – out of necessity – to have each person be a generalist. Clearly, if you only have two or three people they are going to need to cover for each other while one is on vacation or on break, etc.

Specialization is the next step as your team grows. Specialization helps to improve service levels if your team is split up into skill domains. For example, you can have a team that takes care of billing issues, a team that takes care of complaints, a team that focuses on service issues, and yet another for on-boarding newer customers.

Specialization will require the ability to intelligently filter or route post items from one person to the next depending on the skills the agents possess. It can be complicated, but most modern engagement platforms have a way to achieve this.

You can also take the approach of using some care agents who possess multiple skills and also have "escalation" specialists in the mix.

A hybrid approach like this is effective as you grow your team or are transitioning from generalist to specialist team make-up.

Approach

Your approach strategy is likely to be influenced by your company's management style and philosophy regarding customer care. For example, some customer care centers focus on the quality of the interaction with each customer – even if that means longer queue times and longer progress times. Sometimes cost savings and efficiency initiatives get in the way of this idea. In other words, if you are getting messages from management that you need to cut average handle time in half, it is hard to concentrate on quality.

Getting this out in the open is important in cementing your strategy. No one needs a strategy that cannot be put into good practice.

A middle ground is to establish and maintain well-defined business rules for each situation. These business rules give guidance to supervisors and agents alike and can supersede the notion of volume vs. quality. For example, there may be a business rule that dictates: "Any customer who, within the first thirty days experiences a product problem and is angry as a result, shall receive a $15 credit on his next bill."

Rules like this can help to streamline interactions and help to define "quality" without having to make a one-off decision each time there is an angry customer.

A volume focus is a different strategy altogether. Here, the idea is to process as many interactions each hour as can possibly be done. This means quicker assignments, fewer dialog turns (e.g. responses and replies), and quicker resolution time. Using this approach, you would be driving your agents to reduce the amount of assignment time, progress time and overall handle time for each transaction.

If you don't know where to start, or if you are new at running a customer care operation, it is a safe bet to take a quality approach. Then, once you have cataloged trends and outcomes of interactions, you can begin to develop best practices around your history with customers. These best practices may later be drafted into business rules. These business rules can be overlaid on top of a quality approach. Then it becomes either a matter of time or a matter of management policy if you migrate into a volume approach.

Content

The way your social care agents create and transmit content back to customers actually requires a strategy. If you intend to establish a "high touch / friendly" social persona, the "pure social" approach to content may be enough.

With pure social content, you are doing a lot of listening and responding. You act as a sounding board. Ditto, you act as a concerned advocate. This friendly approach can be effective in showing how much you care, but it is usually not a good approach if you are offering technical support or solving very specific problems.

KB (knowledge base) referral is another content approach that seeks to educate or inform. A knowledge base is basically comprised of stored answers to frequently asked questions. These can be conveyed with semi-automated agent scripts, or alternately by pushing URLs to social customers inside of a Tweet, Facebook post, or blog reply.

Channel conversion is another way of dealing with content. Channel conversion means "switching channels" to another communications medium. For example, social care agents often have to help customers navigate through very complex problems that simply cannot be answered in 140 characters. When this happens, you can switch channels over to a phone call or maybe a chat. This can be done by pushing a chat or telephone callback URL page to the customer in the form of a private message.

So converting a tweet into a phone call is a form of channel conversion. Same goes for converting a tweet into a chat. The trouble is, some customers may push back on this idea. They may say: "Hey, if I wanted to do a phone call, I would have called." That sentiment has its own challenges, but in the end, a customer care agent's job is to solve problems and answer questions.

You may be in a position where you simply do not have enough manpower to answer each social post item in its native form. You may only have one or two social agents and it's all they can do to pitchfork customers over to a traditional contact center whenever they are able.

Key Drivers

Next, let's take a look at the key drivers for your customer care initiative. First, there's sentiment. If you wish, you can run your care center based on the notion that "the squeaky wheel gets the oil." This sentiment driver is the basis for escalation protocols wherein the angrier customers are handed-off to specialists skilled at handling folks who are not happy. Even if you do not have a large enough staff to do tiered escalations, you may nonetheless wish to base your outreach efforts on sentiment.

This is where the exercise of detailing the sentiment and influence across themes comes in handy. Let's say you have done a basic sentiment analysis and the majority of the post items you see are neutral or happy – with only a few outliers that are unhappy.

In that case, you may decide that basing your outreach efforts on sentiment is not a great idea. If the opposite is true, and more customers are unhappy, this may be the right approach.

Influence can be another key driver for outreach strategy. But as we will discuss later, influence can be tricky. For example, do you want to focus entirely on public influence as in the Klout score of an individual? Or do you want to look at how influential a customer is in terms of their brand loyalty? (years as a customer or annual spend – also called corporate influence).

Some social care experts insist that you must look at both types of influence – public and corporate. This gives you a more rounded view of what's going on with the customer. Most modern social engagement platforms tout built-in public influence scoring, but corporate influence is still limited to a few advanced platforms.

The more advanced systems give you the ability to do CRM or database integrations so you can glean corporate influence as well.

The type of platform you choose and deploy may be a determining factor in whether or not you choose influence as the key driver for your engagement strategy.

In addition to sentiment and influence, you can also rally strictly around customer loyalty. That is to say you can "tier" your customers based on their status as a "silver, gold, or platinum" customer. This strategy requires either knowledge of or integration to customer record systems.

This is possible with off-the-shelf CRM systems or even proprietary customer database software. Either way, you will be mobilizing you team around how "important" the customer is in terms of loyalty and lifetime value.

Supervision

Supervision strategy is often a mirror of the management style of your enterprise. Either that or your social care center will be re-defining culture. It is often the case that social care teams buck tradition and do things their own way. Regardless, the style of supervision you will employ is an important and strategic decision.

In a "tight control" scenario, you will be establishing and monitoring agent KPIs (Key Performance Indicators) and rewarding or disciplining you agents based on adherence to KPI goals.

For example, the number of post items outreached (closed) per shift, and the number of post items assigned, and the average and maximum progress and handle times for interactions. (These metrics and others are addressed in Chapter 13).

In a "Quarterbacking" scenario you will have one or more floating coaches (quarterbacks) who will preview post items, prioritize them, and provide suggestions to agents as to whom will take certain posts. This can be done semi-automatically by "binning" post items into buckets, or it can be done manually depending on volume. A Quarterback "calls the ball" as it were, and provides guidance to the rest of the team. The emphasis here is on collaboration.

Agent Autonomy is another choice altogether. In this scenario, each agent has his or her own goals and work style, but they pretty much call their own ball.

If you have a handful of highly skilled and trustworthy social agents, this is possible. But this approach is not recommended for larger teams.

Metrics

The final part of the "Strategy Bingo" exercise deals with metrics. Simply put, these are data used to measure the productivity and effectiveness of your social engagement initiative. Fortunately we can borrow heavily from established customer care best practices to get moving in the right direction.

Group SLAs are one approach and you can mix and match the metrics to suit your organization. A group SLA measures the overall effectiveness of your agent groups. You may have a billing group, a service group and a retention group for example. The idea with a group SLA is to establish meaningful metrics that allow you to track the overall effectiveness of a single group over time.

For example, you can track the average and maximum handle times for post item transactions. This gives you a handle on how long it takes to see a post item, and then deal with it, do outreach and resolve it.

It's a good idea to see the "maximum" times in this example because it represents the "worst case scenario" for that time period.

Agent KPIs are another approach to metrics wherein the focus is on individual productivity. For example, you can track metrics like queue time (how long it takes for an agent to take a social post item out of queue); and progress time (how long it takes to work on a social post item before final resolution), and other performance indicators such as total time available during a shift.

Disposition trends give you a handle on workflow progression. For example, you can track the number of current (final) dispositions in a specific interval. These may include the number of items labeled as "Outreach Completed," or "Transferred to Sales," or "Outreach in Progress."

Disposition trends are useful in understanding the distribution of workload – that is to say where all the work is or if there are certain individuals where post items are getting hung up. These types of measurements help you as a manager or supervisor to re-distribute work or to move people between different groups.

It could be argued: "All of these metrics we've discussed are important, so why not use each one?"

Once you get some experience with these metrics you can do that, but to get started you should focus on the most important things first and then add on complexity as your team settles in.

Whether you've used 3.5 cards on a table, a whiteboard, or some other method to track your strategy choices on the "bingo card," you should be at a place after some discussion to circle the ones you feel most comfortable with. Your card may look something like the sample below...

Table: Completed Strategy Bingo Card

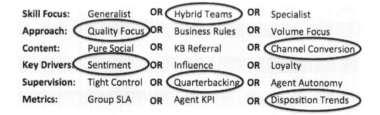

Skill Focus:	Generalist	OR	Hybrid Teams	OR	Specialist
Approach:	Quality Focus	OR	Business Rules	OR	Volume Focus
Content:	Pure Social	OR	KB Referral	OR	Channel Conversion
Key Drivers:	Sentiment	OR	Influence	OR	Loyalty
Supervision:	Tight Control	OR	Quarterbacking	OR	Agent Autonomy
Metrics:	Group SLA	OR	Agent KPI	OR	Disposition Trends

No matter how your bingo card looks exactly you owe yourself praise. Why? Because you've completed perhaps the most important part of your journey in setting up your social engagement for customer care initiative.

You've made keen observations about what customers in general are saying about your brand. You've drawn important conclusions about these observations. Finally you've developed an interim strategy for skill focus, approach, content, drivers, supervision, and metrics.

Now you're ready to take the next step in conducting a thorough discovery process. This next chapter on discovery will take you through the essentials steps. You'll continue on your journey armed with a well-formed strategy. Let's continue!

2. Essential Discovery Steps

The discovery process can be fun but it's also challenging because there are a lot of details you'll have to document. Don't do discovery in a vacuum, but rather get all of the appropriate stakeholders involved.

For example, if you are in charge of customer care, you should seek to involve the supervisors of the social care team, a few social care agents, and a business analyst that is familiar with your procedures and metrics.

This chapter provides you with some simple tables that you can use as worksheets for your discovery process. You can use electronic spreadsheets and distribute them to the folks helping you with this project, or you can mock these up as large whiteboard templates. You can also use big poster sheets on an easel if you are able to meet together in a conference room.

The purpose of conducting the discovery process is to prepare for the "provisioning" of your social engagement for customer care platform. Each of these commercial platforms has some kind of administrative interface into which important data are entered to get your system up and running.

If you follow the advice of this chapter, you will have all of the essential data you need to make the provisioning process swift and efficient. The cleaner and more complete your discovery process, the better the overall implementation will be.

Profile Building Assumptions

An essential part of the discovery process is the creation of profiles. First, let's understand what a profile is so there is no confusion.

A profile is like a "filter" for finding post items on the social web that deal with a specific topic or idea. A profile may represent one of your company's brands. Or it may represent a concept such as "I need help" or "ABC Company is horrible because...." Profiles are important to understand and plan for. This is because the profiles you create define the basic theme and content of the social post items your system is going to fetch. The result is what's served up to the agents on your social care team.

Profiles are defined by key word searches. The key words and phrases that you set up as your "social search" criteria are kind of like a dial on a radio – allowing you to "tune in" to social posts dealing with a certain subject or idea.

Boolean operands such as "OR" and "AND" and "NOT" are used to create search formulas that make-up your profile. For example, the phrases: [("ABC Groceries") AND (Fruit OR Vegetables)] will be used to fetch social post items that include those phrases – all dealing with a fictitious company called ABC Groceries.

Depending on the vendor, your choice of key words and phrases (and the number of profiles you build) may have an impact on what you pay each month to keep your system up and running. So it's important to choose the right profile strategy so you are not paying too much.

On the flip side, not garnering enough social posts may hamper your ability to outreach to customers who really need your help. It may take a little experimentation to strike the correct balance. This experimentation is normal when you set up new profiles.

You can get started by creating a simple list or spreadsheet that contains a name placeholder for each profile you are contemplating. You can then take your best guess at the phrases or words you want to use to define that profile.

It's typical that your social engagement for customer care vendor will build the actual queries, but to get started you should jot down the general themes you desire to "tune in" to.

Table: Sample Profile Worksheet

Profile Name	General Words	Focus Words
Fruit	ABC Grocery, Fruit, Vegetables,	Spoiled, Rotten, Wilted, Fresh, "Not Fresh"
Vegetables	ABC Grocery, Vegetables,	Spoiled, Rotten, Wilted, Fresh, "Not Fresh"
Meat	ABC Grocery, Meat, Poultry, Fish	Fatty, Expensive, Old, Rotten, Spoiled
etc.	etc.	etc.

Static Clusters

Static clusters are "like issues" based on key words associated with the Profiles used to track the progress of outreach to customers and authors.

A static cluster is neither "open" nor "closed", but rather a way of tagging media stream items to indicate a particular item fits a topic associated with a Profile. It is important to distinguish static clusters from trending topics. This is because you can use static clusters representing general business issues that are *persistent*. On the other hand, trending clusters (sometimes called conversation topics) are more of a subtext or conversational undercurrent that can be associated with a persistent business issue.

The two can be used together as I will explain in Chapter 7 on multi-tiered clustering. But or now, let's concentrate on static clusters.

Here is a sample grid you can use to establish the static clusters you will be using that are associated with the profiles you are setting up:

Table: Sample Static Cluster Worksheet

Static Cluster Name	Agent Skill Required	Department or Specialty
Service Issue	Engagement, escalation	Customer Service
Coupon Related	Engagement, research, confirmation	Sales
Returns	Engagement, policy	Customer Service
Deliveries	Engagement, follow-up	Dispatch
Sale Items	Engagement, research, confirmation	Sales
Rant	Engagement	Customer Service
Miscellaneous	Quarterback, research	Tier 2 Supervisor
etc.	etc.	etc.

You can create a default "miscellaneous" cluster to be assigned to all social posts that do not fit any of the pre-defined static clusters.

Social Token Data

Modern social engagement platforms allow you to establish multiple proxies so many agents can share a single Twitter or Facebook account and respond from that same address. A proxy must be authenticated against the established Twitter and Facebook accounts you intend to use.

In this authentication process, an administrative console is used to direct the administrator to a special Twitter or Facebook-hosted web page. This happens after the appropriate credentials are entered. Once this is done, the social network that's authenticating the application will transmit a token. This token is stored in your social engagement platform. The token establishes the application as a "trusted" application. Now, your system may converse with your Twitter and Facebook accounts as if you were logged in and conversing natively on your corporate Twitter page or Facebook page.

This proxy is necessary so all of your agents can share the same account in responding to social post items. (In professional social engagement platforms, agents respond from a corporate account, not using their own persona).

Depending on the number of brands or profiles you are managing, you may be using more than one Twitter and/or Facebook account. For example, you may have a Twitter account set up exclusively for customer service and another one for technical support. Ditto Facebook, where you may have various fan pages set up.

Table: Sample Social Proxy Token Worksheet

Profile	Network	User Name	Password
Produce	Twitter (#1)	@abcmarket	***rufjts
Produce	Facebook (#1)	/abcmarket	***76yt6
Meats	Twitter (#2)	@abcmeats	***uythg8
Meats	Facebook (#2)	/abcmeats	***itujf8
Deli	Twitter (#1)	@abcmarket	***rufjts
Deli	Facebook (#1)	/abcmarket	***76yt6
Customer Service	Facebook (#3)	/abcservice	***puy6t5
etc.	etc.	etc.	etc.

Roles and Permissions

When you set up the infrastructure for your social engagement team, you will need to establish roles for each person on the team. The roles you define each have a set of permissions associated with them. These permissions govern what functions each person has access to.

(In most social engagement platforms, the role you play equates to specific permissions or "class of service" that needs to be entered into the system using an administrative interface).

During the discovery process, you can keep track of all these roles using a simple spreadsheet or table.

It's not a bad idea to memorialize this outside of your engagement platform so no matter what platform(s) you are using, you always have this information at your fingertips.

Table: Sample Roles Worksheet

Name	Email	Role
Sally Proctor	sproctor@abc.com	Listen
Bill Hyde	bhyde@abc.com	Engage
Jasmine Neet	jneet@abc.com	Supervisor
Ed Nystor	enystor@abc.com	Engage
Frank Blythe	fblythe@abc.com	Engage
Gita Mimba	gmimba@abc.com	Engage
Tara Nice	tnice@abc.com	Engage
Mitzy Minz	mminz@abc.com	Listen
etc.	etc.	etc.

Let's talk about the different roles team members will be playing. First, there is a "listen-only role." A person who plays this role is typically a business analyst, brand manager or marketing person.

These folks need to see what's going on (trends, sentiment, top voices, etc.), but they do not interact or engage with authors inside of the engagement platform.

Second, you will of course have people playing the "engage role." These are social care specialists who are authorized to do outreach to authors and respond to their posts.

Next is the "supervisor role." Supervisors often have the ability to manipulate which agents see certain types of posts. They also have the ability to review individual agent KPI statistics, and group SLA statistics. There is also the "social quarterback" role.

A quarterback can play the supervisor and the engage role. These are often specialist who can "call the ball" on what posts should get special treatment, what trends are worth investigating, and strategies for re-distributing the workload.

The fourth role is the "administrator role." This person is in charge of making configuration changes. For example, changes to key word search criteria for profiles, or the grooming of trending topics (word clouds), or the provisioning of new social network proxies. Administrators also do moves, adds, changes and deletions of user accounts and set the roles and permissions for each user.

Groups

Groups allow you to create discrete teams of users for routing and reporting purposes. For example, posts for a certain profile can be assigned to a specific group.

Reports can also be filtered by agent group. This helps supervisors and quarterbacks to monitor metrics for specific groups rather than the entire agent population.

Table: Agent Group Worksheet

Name	Support	Marketing	Service	Retention
Sally Proctor		X		
Bill Hyde	X			
Jasmine Neet	X		X	X
Ed Nystor	X			
Frank Blythe			X	X
Gita Mimba			X	X
Tara Nice	X			
Mitzy Minz		X		
etc.	etc.	etc.	etc.	etc.

It's a good idea to develop talent amongst your agents so they can cross over into more than one discipline. This is more important with smaller teams than it is with larger ones. You can keep a tally of skills and skill proficiency as a tool to choose agents for each group. I recommend first listing the skills, and then rating them for each person on a scale of one to ten.

Here is an example of the kinds of skills you can keep track of for each agent:

- Problem resolution
- Communication Skills
- Technical expertise (subject matter expert)
- Customer Service Escalation
- Fluent: French
- Fluent: English
- Fluent: Spanish
- Risk Mitigation
- Customer Retention
- Public Relations
- Policies and Procedures
- Management skills

Once you have this list for each agent and have noted their skill proficiency, revisit the list once a quarter to see if any of the skills have changed or new ones have been developed. This will help you to form and make changes to groups wisely.

Dispositions

Dispositions allow agents and supervisors to label the status or outcome of each social post item they work on. If you are new to social engagement for customer care, there are some nuances about dispositions that may not be readily apparent. For example, let's say you are the first agent to take a social post item from the queue.

Next you read it and determine it could be handled better by someone else. In this case, you may change the disposition to "Transferred – Tier 2 support."

Once you select one of those transferred dispositions, there are two ways the correct person could get the item into their personal workspace.

First, the transferred disposition itself can be used as an "ACD" (Automatic Call Distributor) trigger. This means that a universal queue management engine could scan for those transfers and use that as a means to "push" that item into the correctly skilled agent's workspace.

The second way the correct person could get an item that you disposition as "Transferred" is for that person to filter on that transfer name. By using advanced filtering, agents can concentrate on social post items that are "Transferred – Tier 2 Support" or "Transferred to Retention Desk," etc.

Taking this to the next level, the most advanced systems allow you to sort post items by spam score, sentiment, profile, cluster and other attributes.

You might be thinking: "Wow that's a lot of power to put into the hands of an agent – where is the supervisor in all of this?" You are of course right to think this! More advanced systems have the capability for Supervisors to "lock down" the filtering on an agent-by-agent basis. This is called Supervisory Filtering or Lock-Down Filtering.

There are three basic disposition states. They are: a) Open; b) Transferred, and c) Closed. Each of these states can have more than one disposition type. Here are some sample dispositions to get you started:

Table: Sample Dispositions

State	Disposition Name
Open	Researching Item
Open	Outreach in Progress
Transferred	Transferred to Tier 2 Support
Transferred	Transferred to Sales
Transferred	Transferred to Retention Desk
Closed	Outreach Completed / Resolved
Closed	Not Applicable

Agent Presence Settings

Agent presence settings are used by social engagement platforms to establish the availability and work habits of agents. There are three basic reasons for establishing the presence and availability of an agent:

- Helps in routing so social post items don't languish
- Helps supervisors to manage their teams
- Aids in establishing work hours for reporting

A typical presence management table for supervisors to view looks like this:

Table: Agent Presence Settings

Agent	Offline	On Break	Available	Busy
Sally Proctor	X			
Bill Hyde		X		
Jasmine Neet			X	
Ed Nystor	X			
Frank Blythe				X
Gita Mimba				X
Tara Nice		X		
Mitzy Minz			X	
etc.	etc.	etc.	etc.	etc.

It's mostly up to the supervisors and management team of your company to decide how granular you want to get in terms of presence reporting. For example, you might take the basic route and just stick with "available" and "not available."

That will work for ACD or automated routing purposes, but it does not give you a lot of insight into agent occupancy (effective work time of the agent).

You may want to gather statistics on workflow and work habits as well. For example, you may want to establish presence states for "wrap up," "idle time," and "research."

Wrap-up activities include back-end fulfillment, order processing, trouble ticket paperwork and other work that happens after the transaction with the customer. It's all legitimate work, but the time these tasks take should not count against the effective work time (occupancy) of the agent.

Idle time is time that the agent is doing nothing productive – just waiting around for something to happen. This is not productive time and counts against agent occupancy.

41

"Research" may seem like idle time, but it deals with any activity that the agent must do in order to be more responsive to customers. For example, you may hand out news briefs, product data sheets, press releases, etc. to your agents so they can learn about issues that will aid them in their outreach activities.

Congratulations! You've just spent a lot of time and effort to collect all of the essential data you'll need. Now the provisioning of your social engagement for customer care platform can commence. You can rest assured that this information will make it a lot smoother in getting your system up and running.

One warning though – and that's "design drift." This is how a finely tuned system can drift away from its optimal settings over time. This happens because new initiatives arise; new agents with new skills join the team, products and offerings change, etc. It is therefore necessary to revisit the strategic and discovery processes with frequency. The effort you invest here is well worth it. Otherwise your system will not be working up to snuff and neither will your team.

3. Agent Assistance

You can ensure that your agents are fully prepared and ready for social customer outreach by taking the time to develop proper agent assistance tools for them.

It helps to start with a social engagement platform that is specifically designed for customer service agents and supervisors. But beyond this, you can give your team that extra edge by following the advice in this chapter.

There are three fundamental sets of tools that provide agents with assistance in a social customer care setting.

First, priority tagging gives your agents the ability to quickly recognize and react to social post items in the form of banners, colored icons and tags. These tags draw attention to specific attributes of a post. For example, how old the post is, or whether or not there is an existing conversation thread going on with the author of the post.

Second, agent assistance scripts provide ready-made answers and suggestions for customers. You can catalog these and name them easily. Modern engagement platforms give you the ability to pull these up and automatically populate the response pane with this text.

Third, next best actions (NBA) provide special instructions. These instructions are meant to be agent-facing and not customer-facing. Next best actions are triggered when certain, pre-determined conditions exist. All three of these agent assistance tools are covered in more detail below.

Priority Tagging

Modern social engagement for customer care platforms have the ability to "tag" social posts based on specific attributes. For example, the post of an angry blogger who has a high social influence score could be tagged as a high priority. This is called priority tagging.

Depending on the platform you are using, there are various attributes of a social post item that can be used to trigger priorities in collaboration with a rules engine. Some examples of these are:

- The Source URL and title of the post item
- Selected text in the post body
- When the post item was published
- Source (social media, blog, article)
- Author name or handle
- Author public influence
- Author corporate influence
- Post sentiment
- Persistent Cluster (Business Issue)
- Trending Cluster (Trending Topic)

Here is an example rules sheet you can use to start building the conditions and triggers that will provide powerful agent assistance for your team:

Table: Sample Priority Rules Worksheet

Profile	Tag Name	Conditions	Priority
Fruit	Priority Blogger	1. Blogger 2. Influence over 30% 3. Negative Sentiment	2
Fruit	Priority Customer	1. Social Post 2. Corporate Influence +20 3. No answer in 24 hours	1
Meat	Priority Article	1. Local Article 2. "Spoiled" Mention 3. Negative Sentiment	2
Meat	Priority Customer	1. Social Post 2. Corporate Influence +20 3. Negative Sentiment	2
Coupons	Coupon Problem	1. Social Post 2. Coupon Cluster 3. "Rain Check" in body	3
etc.	etc.	etc.	etc.

If you do not have a platform with a built-in rules engine with which to trigger these priority tags automatically, you can nonetheless prepare your agents by providing them with a "cheat sheet" that can be posted next to their workstation or on a large poster on the wall.

The important thing is to establish the conditions that help guide agents on the priority of posts. This will make them much happier because their workday will be more productive, and productive team members make for happy team members.

Agent Assistance Scripts

Agent assistance scripts are pre-defined (canned) textual messages agents can use to respond to social post items. In most cases, these scripts are editable so agents can personalize them a little for each customer. You can also use agent assistance scripts to pass URL links. These links make it easy for customers to view information that is longer than 140 characters (the Twitter character limitation).

Depending on your internal policies, your legal department or management team may need to approve these scripts. This may be true of financial services or healthcare companies, for example.

Modern social engagement platforms provide you with the infrastructure to capture and catalog these scripts. Some also allow you to automatically populate your "response pane" by clicking on a script name.

This saves steps and keystrokes for your agents and makes compliance a lot easier to implement.

Even though agent assistance scripts are typically canned, it is a best practice to make them editable so agents can personalize them before sending them out. This personalization step helps to avoid the appearance of the responses being "robotized."

That is not to say that robotic responses are bad. But it's best to limit robotic responses to social handles that are well known by your customers as being robots, and not attended by human agents.

Next Best Actions

Next best actions (NBA) are explicit instructions telling agents what to do based on specific conditions. These conditions may be similar to or exactly the same as priority tags, but in more advanced systems, they work independently. Not all engagement platforms have this advanced capability, but I recommend you investigate this fully.

Table: Sample Agent Assistance Worksheet

Profile	Script Name	Script Shorthand
Produce	Keep Fresh Tips	Here is a great article on keeping fruit fresh when you get home: bit.ly/7586ty
Produce	When is Fruit Ripe?	Some tips on knowing when popular fruits are ripe: bit.ly/78urj6
Deli	Weekend Specials	See our coupon section on the web site acme.com/coupon
Meats	A word on fat content	Great post on understanding fat content in hamburger: bit.ly/98isu7
etc.	etc.	etc.

There are several benefits of using NBA scripts. First, they are great for on-boarding new agents who are not familiar with all of your workflow rules and policies. For example, you can set up an NBA on the conditions that an influential customer is angry about a specific product.

That's three attributes: a) Customer; b) Angry Sentiment; and c) Product "X." The NBA can be triggered when these specific conditions exist.

The script may say, for example: "Influential and Angry Meat Department Customer – escalate this to the Retention Desk Immediately." Likewise you can push an NBA script that gives guidance on refunds, repairs, replacements – just about anything that you explicitly set down rules for.

Next best actions are more powerful when your social engagement for customer care platform is linked to a CRM (Customer Relationship Management) system. This allows for the incorporation of corporate influence scores, loyalty rankings, and other essential customer information that can be used to trigger NBAs.

This agent assistance tool is also effective in agent adherence to compliance issues. For example, some agents are required to recite certain scripts based on the type of transaction.

For example, "Are you over the age of 18 and responsible for this bill?" For an agent to be in compliance, he or she must say (or write) such a phrase if certain conditions exist. The same applies for social engagement.

You may decide that for your organization, you always want to send a loyalty coupon or offer some kind of incentive to certain customers who have experienced something bad with your brand. The possibilities are endless. The best starting point for these NBAs is your own company's workflow and policies documents. If these policies are not written down, I urge you to do so. Memorializing your policies in the form of triggers and next best actions will save you many steps in agent training and coaching, and also possibly keep you out of big trouble.

Table: Sample Next Best Action Worksheet

Profile	NBA Tag & Triggers	NBA Instructions
Produce	(PR Escalation) Priority Blogger, Negative, +40 Influence	Please escalate this post to the PR department for a customized response
Produce	(Loyalty Priority) Social Post, "Refund" in text, Negative sentiment, any corporate influence	Offer this person a coupon for redemption at next visit. Offer code #655
Meats	(Loyalty Priority) Social Post, "Spoiled" in text, Negative sentiment, any corporate influence	Inform District Manager, send quality note #455, offer coupon code #354
etc.	etc.	etc.

As with agent assistance scripts, next best action scripts are native tools in advanced social engagement for customer care platforms. If you are using a platform that does not have built-in NBAs, you can still keep track of these actions on a table that you can publish to your agents. You can start with cheat sheets, posters, and wallboard displays and then graduate to an engagement platform that incorporates all of these.

Agent assistance tools are essential for superior onboarding, agent retention, and agent adherence to compliance rules. Make a habit of reviewing your workflow rules and policies and publishing them to your team. Better yet, do quarterly reviews and integrate your agent assistance tools into your engagement platform for superior customer care.

4. Engagement Etiquette

Social engagement for customer care is perhaps one of the most exciting and growing areas for customer service professionals. More and more, consumers look to their social networks for help, advice, buying tips and even technical support.

But you've got to be careful how you outreach to social authors because you don't want to be characterized as a "cyber stalker" or just plain rude. Here are some etiquette tips being put in to practice by social care experts.

Open Ask vs. Private Mention

Despite the fact that Twitter and Facebook are open forums, many people use these social networks as private mediums and conduct one-on-one conversations in public view. Of course there is a more appropriate way to communicate privately: direct messages. Despite this solution, many nonetheless converse privately on an open channel.

The question is, does that give you the right to directly answer a private question even though you were only eavesdropping? Well, the answer is not really. You run the risk of alienating your target audience and hurting your brand. This means if you send a tweet with an @person handle to get that person's attention, it may be too direct.

A less direct approach is to first "follow" or "friend" the person you think you can help. The person you are reaching out to will accept your friend request or follow you back if they are interested in getting your advice. Of course it helps to have a handle that is obvious so the person you are interested in communicating with understands which brand you represent.

For example: @abccompanysupport or @abccompanyhelp are obvious company handles.

On the other hand, if someone is asking an open, "croudsource-like" question that is not directed at a specific person, it is fair game to answer with an @handle response. That's not cyberstalking, but simply an outreach based on a "public" question in the form of an open ask.

Understanding the difference between an open ask and a private mention is important so you don't step on people's toes or get them upset. Read each post carefully so you approach it appropriately.

Offer Privacy

It's a good idea to offer an option of privacy when responding to social posts. For example, if someone is letting off steam about rotten fruit, you can say: **"@person would love to help. Please follow me and I'll help with a direct message or phone call."**

If the person does not want to follow you a direct (private) message is not possible.

A way around this problem is to have a knowledge base set up with published articles answering questions or giving tips on popular subjects. You can tweet or post the URL associated with these articles.

For example, if you have an article prepared that explains the procedure for returning a products and getting a refund, you could tweet this publicly: "@person returning items is a snap. Just go to http://www. returns.abc.com/art443 to learn how to get a refund." Clearly you don't want to convey private data over a public medium but sending a URL to generic knowledge base articles is OK.

Be Level Headed and Neutral

As a social engagement care professional, the persona you develop on line is more of a corporate persona, not a personal one.

You are speaking on behalf of your company, and therefore you need to tune-in to the corporate persona, not your own. Each company has its own guidelines for how to engage and how to deflect angry posts. Some simply ignore rants and look for more neutral or less emotional posts. Some engage regardless of sentiment.

Either way, it is a best practice to remain neutral and calm when responding to an author's posts. Breaking it down:

1. Don't use emoticons or capital letters
2. Don't judge or talk down to people
3. Always offer help: "I think I can help..."
4. Offer an escape valve: "Let's Chat when you get a moment."

Develop Rules and Next Best Actions

It's a good idea to tie responses to pre-developed scripts and "next best actions." For example, if an angry customer has had two "lemon" experiences in a row with your product, management may allow a "no questions asked" refund or a coupon or discount on the next order. You will need to develop rules for each type of situation and make those rules available to your social engagement care team.

Once you've developed standard rules, you can tie each rule to a specific script. Now your agents will have an "official" response for the most likely scenarios.

You can give them the flexibility to change them up a bit so they don't sound the same all the time. The last thing you want is to give the impression that the responses are automated – unless that's the expectation you have set.

By following a few best practices such as offering privacy, non-threatening generic answers, and staying neutral, you can navigate social networks as a social engagement care professional without getting yourself or your brand in trouble. It is helpful to draw-up your own rules of engagement for your company. This way, everyone involved in social engagement for customer care is using the same playbook.

5. Getting a Handle on Social Influence

There are all kinds of metrics we use to quantify the importance of a social post item and social authors. Amongst them is topic relevancy and sentiment. But perhaps the least understood or leveraged is social influence.

Let's take a closer look at influence and see what can be done to make it more useful.

What is Social Influence?

In the context of social engagement for customer care, social influence is an attempt to score or rate how influential an author is. Influence scoring is available via social feed aggregators and also companies such as Klout, for example. Numerous algorithms are used to calculate an influence score.

Some of the things that have a bearing on a social influence score are: a) the number of followers, connections, or "friends" a person has; b) the number of posts a person has; c) the frequency of posts; d) the number of views or hits on a blog or article landing page; d) number of "likes" or "pins."

Curb Your Enthusiasm

Well, don't get too excited just yet. Yes, you can use social influence scores as a data point you need in making outreach decisions, but it's just a data point. First, let's take a look at some things you should consider:

1. Influence Gaming. Social networking users can "game" their influence by exploiting "junk follows." This is called influence gaming. A junk follow is a throw-away follow based on marketers, scammers, and robots that stalk you just to get their own follows for commercial reasons. "I am following you now, so follow me back and I will show you how to get one million followers." By indiscriminately following these solicitations, which turn propagate other reciprocal follows, you can "game" or goose-up your influence ratings. You can even pay for fake followers and even re-tweeters if you want professional help in gaming your influence score.

2. Domain Expertise. Most influence scoring algorithms do not take into account the author's actual domain expertise. That is to say that influence can be "watered down" if the author in question is posting on random subjects and is not recognized as a thought leader on a certain subject. This would include the celebrity quotient - wherein people have a lot of followers because they are a celebrity, not because they are expert on a certain subject.

Clearly, it is important to know if someone tweeting about your brand is a celebrity with a high influence score, but it is also important to consider his or her impact.

For example, consider a tweet from a celebrity not recognized by the constituency you are targeting. If your constituency does not recognize that celebrity, his or her tweet may not be very important.

3. Relevancy. The influence of an individual may not be important if their mention of your brand, hash tag, etc. is out of context from a customer care perspective. If the mention is out of context, it is probably not relevant. If your name or brand is simply mentioned, but there is no sentiment or intent associated with it, it may not matter how influential the poster is.

Is there more than Social Influence?

There are other kinds of influence that can be important from a customer care perspective. Consider the fact that in a customer care setting, influence goes beyond the number of followers someone has. It goes deeper.

For example, your company may have a commercial relationship with the author. That makes the author a customer. You are probably tracking the loyalty, buying patterns, revenue spend, and history of that customer. Such is the stuff of a "corporate influence" score.

Here, you can use your own algorithm to determine how important a customer is to your enterprise. Modern social engagement for customer care systems, especially if they are CRM-integrated, can take advantage of corporate influence scores.

Imagine being able to see public influence and corporate influence scores side-by-side and sorted by author. This comparison can help social care agents make quick decisions as posts by those customers are shown.

Automating on Influence

Modern social engagement for customer care platforms will also allow you to define rules by which automated filtering and dispositioning can occur. For example, you might want to set up a rule that selects posts based on the following criteria:

1. Public Influence of over 20 points
2. Corporate Influence of over 15 points
3. "Not Happy" sentiment-wise
4. In the "Lemon Product" trending cluster

Now, you can set up next best actions or automatic dispositions based on these criteria. The list of how you can set up rules using influence is endless.

Tweaking it for your enterprise can be done pretty easily based on the observations your social care team is making and also of course your established business rules. In this context, public social influence scoring can be an effective tool in social care outreach.

These social influence scores are much more effective when combined with other author and customer attributes such as corporate influence, clusters, and sentiment. It's a good idea to explore ways to hook-in customer scoring based on your brand's relationship with the customer so you can go deeper and establish more impactful outreach.

6. Sentiment Approaches

Sentiment is an attribute of social posts that provide rich clues on how to engage with customers. But it's not always what it seems on the surface.

Here are some tips for figuring out the difference between a plain rant versus a cry for help.

Rants and Pans

Rants are important because they embody the spirit of entitlement that is one of the key tenants of social networking. If you want to rant in view of a social network, you have an instant audience and a lot of people will be listening. From a customer care perspective, it's important to "follow" the rants of individuals, especially if they are influential in your commercial space.

Decide to look for trends in the rants of Authors. Does a specific author rant constantly and could therefore be characterized as a serial curmudgeon? If so, don't take the rants too much to heart. In fact serial ranting is a trademark of some influential bloggers.

It's important to separate a "blind rant" from a sincere cry for help. For example a customer may say: "I hate ABC. They are just the worst. No one should do business with them." This rant does not contain any useful information with which to identify an action to take. That is to say: If the post is not actionable, it is useless from a customer care perspective.

Alternately, some rants can be very specific. For example: "ABC just has the worst service. When I call I wait a long time. I was one day out of warranty and they would not fix my problem. I hate ABC." Now, depending on the business rules and exceptions that can be made for customers, this post example is actually actionable and more likely to qualify for agent outreach.

I recommend drawing a line between blind rants and more specific ones. Depending on the policies of your company that line could move a little – but what's important is giving your agents proper guidance on what to work on and what to ignore.

In some cases, you may want to peel-off blind rants into their own work queue so someone who is expert in dealing with really angry customers can disposition them.

Cries for Help, Oblique or Otherwise

Some social posts that cry out for customer care possess a sense of profound frustration or sadness. These are good to keep on your radar screen as a customer care team because they are kind of an "early warning system." For example, someone may say:

"I am just getting so frustrated with ABC Markets. I've made three trips to the market this week and I just can't seem to get a perfect melon." That's not necessarily a rant, but rather a cry for help. It may be an oblique cry for help if the author of the post does not use the company name as part of a salutation, but nonetheless it is an open cry for help. Since it is fairly specific, the customer care team can focus on the level of frustration and especially the "three times" phrase to ascertain the priority of this post.

It is a best practice to improve on your catalog of next best actions over time. You can make these improvements based on the trending content of social post items with strong sentiment. In addition to sentiment, you should also consider triggering next best actions based on frequency, SLA metrics, and influence.

Dealing with Sarcasm

Sarcasm is a close cousin to ranting because it is used to express contempt or disgust to an intended audience. For example someone might say: "Hey great job ABC Markets. I just love it when my fruit is rotten three weeks in a row…"

If a sarcastic post has specific, actionable information in it, it is still a candidate for outreach. So, just because someone is being sarcastic does not mean they don't need help or don't want help.

It's also important to consider that oftentimes, the persona people use in public to a larger audience can be harsher than if it's a one-on-one conversation. People tend to be less harsh when addressing you directly, and more harsh in a public post. On the other hand, some folks are just plain sarcastic as a rule, but that does not rule them out as a candidate for customer service.

The trick with sarcasm is to read between the lines for the source of the problem and ignore the delivery style. The content's what's important, not the delivery style in most cases.

Closing the Loop on Happy

When someone expresses happiness with your company, it's a good idea to recognize that. You can do that publicly or privately depending on the rules of engagement for your company.

One way of recognizing it is to "follow" or "friend" that individual to let them know you are paying attention to what they are saying and that you care. The decision here is how far to take it. If someone who is currently happy with you sees you followed him or her or put in a friend request, it is customary to follow you back.

If they reciprocate, it's recommended that you send a short, friendly direct message: "Sally, thanks for the praise. We will do what it takes to keep it this way. Let me know if you ever need help." The point here is that you should not just be responding to people who are unhappy, but also reinforce the behavior of people who provide your company with praise.

In fact, you can use friendly feedback as a way to "gamify" interactions. For example, you can track the number of mentions you get from customers and use that as an incentive for them to provide more feedback. You can establish a point system that allows them to redeem rewards.

Likewise, you can instill a spirit of competition between users who help each other as experts by giving "stars" for the number of replies they make on certain subjects.

This encourages community and also recognizes the customers who are helping the community. One way to track this is using a "Top Voices" dashboard or report (Chapter 13 on SLAs and Essential Analytics covers this and other common reports in more detail).

The nuances of sentiment provide important clues on how to engage with customers in the social realm. Whether they are rants, praise, or cries for help, you should develop specific rules and actions for outreaching based on sentiment. By doing this you can establish a consistent, and useful social care persona for your brand.

7. Benefits of Multi-Tiered Clustering

Clustering, or categorizing "like issues," is an effective way to accelerate the outreach and resolution of customer care issues over social channels. By categorizing like issues, your social care team can more easily identify hotspots.

With built-in Natural Language Processing, and a rules-based engine, you can automate routing and next best actions related to these clusters, too. But before you lump like issues into a single dimension, consider the benefits of multi-tiered clustering.

First, What's a cluster?

A cluster is a group of words and phrases that make up a sub-topic of common ideas. In the context of social engagement for customer care, these sub-topics represent the undercurrent of a general conversation shared by many people who are posting socially.

Automatic clustering of like topics makes it a lot easier for your social care team to do outreach. Instead of having to read the entire content of each post, you can just eyeball the cluster name. Across your whole social care team, this can save many hours each day.

Initial Static Cluster Analysis

When you first set up a social engagement for customer care system, one of the important steps is to analyze the "word clouds" representing the clusters that are gathered by the NLP/NLU (Natural Language Processing) Engine. A static cluster is a word cloud representing persistent business issues as opposed to trends.

Modern systems allow you to run the NLP engine against all of the incoming social posts so the engine can start to categorize clusters of like phrases. This is an important step because you can use this initial data set to set up your "first tier" of clusters. The idea is to take a look at the word clouds that have been gathered and group them together into issues that are persistent for your business.

For example, if you are running a service business, you may observe word clouds falling into categories such as "billing," "service plans," "usage," "sales," "complaints," and "technical support." For the most part these issues are static, persistent issues that your business deals with every day. They are not really trends, but common categories you have probably already mobilized a customer care team around.

By way of example, the phrases: "the whole bag was spoiled," "the tomatoes were rotten," and "having trouble keeping produce fresh" all share a common theme of spoilage. Seeing these phrases in the first traunch of word clouds will help you to call this static cluster "freshness."

You can validate what the persistent issues are if you look at the word clouds over days and weeks and they are still the same.

Clusters that show up outside of these persistent issues are for the most part "trending" issues. These trending issues can be used to establish a sub-text or theme in addition to the static clusters.

Trending Issues - The Second Tier

Having a second tier of clusters is useful because you can trigger rules based on the *combination* of persistent issues and associated trending issues. These rules can represent automated actions, or they may represent outreach suggestions that are pushed to agents. In addition, you can do this without disturbing your SLA reporting and analytics that are based on more persistent topics.

For example, you will surely want to report on how many social posts fell into the persistent categories of "billing" or "complaints." But underneath the static cluster of "complaints" there may be certain types of complaints that you don't want your social care team working on.

Let's say, for example, you don't want them working on complaints that are not actionable like: "I hate the color of ABC's logo."

With two-tiered clustering, you could still register the number of static cluster "complaints" but peel-off and disposition as "closed" all those complaints (like the color of your logo) that are not actionable. Or you could even create rules to trigger those complaints as "transferred" to a brand manager or someone else to whom the complaint is actionable.

Other Tier Two Cluster Triggers

Imagine being able to identify a trending topic and then filter it based on sentiment, influence, or how long it's been waiting on a response. That's the power of combining a NLP engine with a business rules engine. Here are some use cases that will help you to think of more for your business:

Trending issue: New store opening and being under-stocked on a certain item. Filter on angry sentiment, and route influential posters into a special work queue. Outreach with a "rain check" certificate, coupon or other offer.

Trending issue: Competitive new product announcement. Filter on happy sentiment, and key words such as "buy" and route to a special "Sales" queue. Outreach with a competitive offering.

Trending issue: Spate of product failures with a common theme. Filter on all sentiment, and key words and phrases: "just bought," "broken," "help," etc. Disposition all social post items from this word cloud into a special service queue. Outreach with an "RMA Hotline" URL.

Two-Tiered clustering is an effective way to approach the categorization; automatic filtering and disposition of social posts for your social care operation.

Take the time to properly characterize persistent, static business issues as your first tier of clusters, so you can take special actions on the trending topics happening around these common business themes. This approach will save your social care team many hours each week and improve customer loyalty at the same time.

8. Profile Strategies

The careful development of profiles for your social engagement for customer care platform has a profound impact on the efficiency of your social care agents. If your profiles are too broad, you get a lot of spam and unwanted noise.

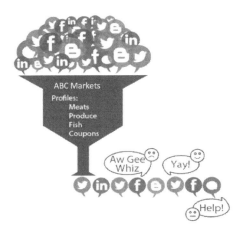

On the other hand, if profiles are too tight, you may miss the opportunity to help a lot of customers. That's why developing profile strategy is crucial. This chapter provides tips on fine-tuning your profiles for optimal performance.

What is a Profile?

A profile is defined by key word searches you set up in your social listening and engagement platform. The key words and phrases that you set up as your "social search" criteria act as a dial on a radio, so you can tune-in to social posts dealing with a certain subject or brand.

Some social care professionals also set up mirror images of these profiles dealing with competitors. This is a way to mine the social cloud for sales opportunities or competitive trends.

Logical Operands

The most common operators for constructing searches are the Boolean operands AND, OR, and NOT. Here below are the "truth tables" that show how these work.

Figure: Truth Table for "AND" (A AND B)

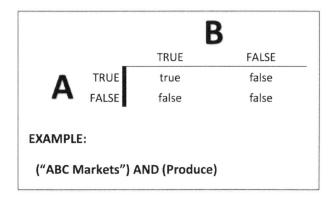

Figure: Truth Table for "OR" (A OR B)

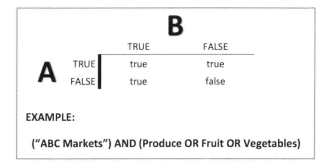

Figure: Truth Table for "NOT" (NOT A)

NOT (NOT A)

A TRUE | false
FALSE | true

EXAMPLE:

("ABC MARKETS" NOT MARKETS) AND (Produce OR Fruit OR Vegetables)

For example, if you run a supermarket chain called ABC Markets, you could set up a profile called: "Produce." An example of what the key words and Boolean operands could be:

("ABC Markets" OR "ABC") AND (Produce OR Fruit OR Vegetables OR Veggies)

With this example, the system would fetch all mentions of ABC or ABC Markets in which any of the words, produce, fruit, vegetables, or veggies was mentioned.

Similarly, you can set up a second profile using key words and phrases dealing with "Meat" or "Delicatessen."

Do a Dry Run First

It is easy to get carried away when you first set up profiles for a social engagement for customer care platform. Some folks have a tendency to set up too many profiles which can make things harder to manage and also more expensive. To avoid this, it is a best practice to do some dry runs first.

For example you can test certain profiles and key word combinations and then simply eyeball the results coming back to see if you are getting too many posts, or not enough.

A good first test is to simply type in the name of your company or brand and see how many posts come back. If you do not get too many mentions, that means your brand is not garnering a lot of buzz, positive or negative. You may then experiment with mentioning more than one brand or even other companies in order to bring in more data.

If you get too many to possibly read, that is a sign that you need to tighten-up the search criteria on the next try.

It is a good idea to start first with broad search criteria. Afterwards, you can incrementally tighten it down on subsequent tries.

Fine-Tuning Keywords

Once you have established the brand and basic subject (i.e. ABC Markets & Produce), you can start to play with qualifying phrases and words. For example, you could create a profile that is searching for social mentions dealing with customer service or quality in the produce department. To do this you might want to add to the search words or phrases like:

… AND (Spoiled OR "not fresh" OR rotten OR wilted OR ruined OR Bad)

By adding this to the original ("ABC Markets" OR "ABC") AND (Produce OR Fruit OR Vegetables OR Veggies) you can create a profile that scans for poor quality produce mentions. This will yield posts in which people are observing or complaining about the quality of the merchandize.

All-in-One vs. Specialized Profiles

It is a good idea to have multiple profiles. This allows you to run reports by issue, product, or brand so you can understand how each one is trending. Some of the reporting metrics we will cover later in Chapter 13 include sentiment, volume and SLA to name a few.

You may want to just do an all-in-one profile that takes in all the key words and phrases you contemplate for your company. But if you take this approach, you will be giving up a lot of control in terms of specialized routing, tagging, and reporting if you do this.

Before you opt for an all-in-one profile strategy, consider three or four broad profiles that all share the same company or brand attributes. For example, one for billing, one for sales, and another for service issues. Alternately, you could profile by product grouping.

Data Feed Diversity

The level of sophistication in developing keyword searches with operators and logical operands is diverse.

For example, DataSift uses its own curation language they call Curated Stream Definition Language (CSDL). This language allows you to filter using three elements: targets, operators, and arguments.

Using this tool, the "target" represents where you want the query to concentrate. For example, a specific character string in a Facebook post, sentiment in a tweet, or a specific language. Of course targets are also sources of data. Depending on your data feed provider these could be Twitter, Facebook, Reddit, YouTube, Blogs, etc.

The "operator" defines the kind of comparison you want done on the data. For example, if you were looking for the word "fruit" the operator would be "contains." Likewise if you were looking for an influence rating, you might use the operator "greater than" on an author's number of followers. There are many operators, including logical operators AND, OR and NOT.

The "argument" is the word or words you are filtering on. For example the target might be twitter, the operator might be AND, and the arguments could be ABC Markets and Fruit.

These targets, operators, and arguments can all be used together to express the profile filter you are creating.

If you wanted to filter on any Twitter posts from self-described consumer advocates who talk about one of several grocery store chains, you could use the following expression:

twitter.text contains_any "ABC, Food Town, Green Grocer, Penn Fruit" AND twitter.user.description contains "consumer advocate."

Another provider, GNIP, provides a rules panel where you type-in keywords for queries followed by options for other attributes such as Klout Score or language detection. Virtually all data feed providers use Boolean logical operands and common operators to construct searches.

Other data feed providers include, TopsyLabs, and InfoChimps. Influence score providers Klout or PeerIndex are typically rolled-up into the data sets of the aforementioned providers.

Once you develop our keyword searches and arguments to create a profile, you will need to tweak them so they are not too "wide open" or too restrictive.

An example of a profile that may be too "wide open" is one that is so broadly defined that you get too many posts - many of which may not be relevant. For example, if you use "Produce OR Vegetables," you will get posts using the word "produce" in various contexts. The word "produce" could mean fruits and vegetables but it could also take in posts dealing with any mentions of the word. Take, for example "I would like to produce a movie" or "every time I ask him to produce a report it is late," etc.

An example of a profile that is too restrictive could be "sentient beings." Such a phrase is not used too often. Ditto if you put geographical or source attributes in your search that limit the reach of the search.

Another challenge in developing keyword searches and arguments deals with using a lot of "AND" logical operands. If you create a search that says: "Produce AND Rotten AND Vegetables AND Tomatoes AND Yellow AND Philadelphia" you'd only get posts that have ALL of those words.

Layering on Rules and Triggers

Each profile you establish drags through hundreds even thousands of social posts. Once you've fine-tuned the key word and search criteria, it's a good idea to characterize your posts as well. Here are some post attributes that can be used as the basis for rule triggers:

- Time of publication
- Source (Blog, Twitter, Facebook, etc.)
- Public Influence Score of Author
- Sentiment of Author
- Cluster (Conversation Topic)
- Existing conversation thread in system

Once you have established these attributes, you can apply them to your rules engine.

For example, you may want to trigger rules on all authors who have an unhappy sentiment who are talking about a hot topic you are tracking. Some platforms will even allow you to automatically route or disposition posts based on certain criteria. If you do not have a rules-based platform, try to at least "tag" the post with these attributes so your social care team can take action based on pre-determined business rules that you set up.

The dry run testing of profile keywords and search criteria will go a long way to making your social care agents more efficient and effective.

Fine-tuning your profiles is an ongoing process but it's worth investing the time so you can get increasingly more accurate and impactful results. It may take a little tinkering but making incremental improvements in your profiles' make-up will ensure a happier, more productive social care team. And a happy, more productive team means a happier customer base.

9. Conversation Threading

Most social customer care professionals find it impossible to track discrete threads of conversation using traditional social engagement tools.

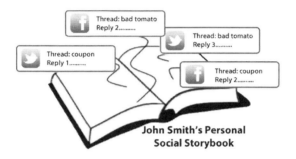

John Smith's Personal
Social Storybook

At first blush, it might not even seem plausible that Twitter or Facebook posts can be manipulated into conversation threads that make sense. However; with the proper tools, not only can you make sense of the jumble, but you can deftly organize threads and track conversations with your social-aware customers.

What is a Thread?

Simply put, a thread is a topic or theme that couples discrete conversations into a continuum. For example, your email client can thread an email conversation with the same subject line. This is useful because at a glance you can see the progression of a conversation and its participants. This helps you to put context on the latest responses and it makes it easier to keep track of who's saying what to whom.

Another way of looking at threaded conversations is to picture a storybook for each customer. The pages will be filled with the text from many different conversations. But each conversation will have a "footnote" indicating the general theme of that conversation.

These themes establish a common thread of conversation throughout the book. Each chapter of the book may have different entries in it – kind of like a diary. And we know that diaries can contain entries that may seem kind of random or disconnected. Themes are picked-up or dropped as you turn the pages.

Now imagine creating this storybook in electronic form. With the use of a database and a program to pull up different conversation themes via computer, you can "browse through the storybook" and look at *all* the themes. Or you can electronically view the storybook with only *one* theme in mind.

The most advanced CRM systems have this electronic storybook capability so agents can "zero in" on specific conversation threads in order to understand what another agent may have said to the customer. This is useful because it avoids agents stepping on each other's toes if more than one social care person is trying to help a customer over an extended period.

Can Tweets be threaded?

Yes tweets can be threaded. Hashtags are an early attempt to do so. It's interesting because I have heard some social experts assert that tweets are meant to be spontaneous and autonomous and therefore ostensibly not connected to a common theme. Some have said that and any attempt to thread conversations is both naive and useless. But the facts prove differently. Tweets can be threaded - if you know what you're doing - and the utility of doing so is a productivity boon for care professionals.

The ability for customer service professionals to search by author and thread on emails is taken for granted. As a care professional, who wouldn't want to be able to zero-in on a specific service issue in seconds instead of stumbling around looking for conversation context manually?

Setting the Stage

Before you jump right in to conversation threading for social care, there are some prerequisites. First, you need the ability to establish an author profile for each person you are having a conversation with. With a CRM-integrated system this is easy. For stand-alone social engagement systems, you of course need a platform that both captures and catalogs each discrete author. The more advanced systems allow you to manage multiple social handles for each author.

Second, your social engagement system needs a database with which to store and pull-up authors' tweets and your responses to them. First generation social engagement systems don't do this so you may need to upgrade.

More advanced systems allow you to scroll through the history of conversations by author - regardless of the social media source of the post.

Third, you need a mechanism to create, store, and recognize conversation threads. Some of the work is up to the agent to determine the moniker of each thread. That's like deciding the subject line of an email, so it's a task most of us are familiar with.

Threading Tips

Once you have the proper infrastructure and social care platform in place, threading strategy is pretty straightforward.

Take a look at the attributes of a specific post to get clues for starting a thread. In chapter 7, we discussed two-tiered clustering. Here, the first cluster is a persistent business issue such as "billing" or "service."

The second cluster is more of a word cloud representing a trending topic. By eyeballing these two attributes, you are able to quickly determine a proper conversation thread name.

For example, let's say the first-tier cluster name is "service" and the trending topic is "long lines." Here the context is how long customers have to wait in line to get service. You could glance at the content of the post and confirm that a good name for the thread may be "long waits at checkout."

Second, if your system allows it, you can tag the post with a "flag" indicating that there is a conversation going on with this author. Based on the rules engine you are using, you may even be able to trigger a flag on incoming posts from that author as a "heads up" there is a current conversation going on with that person. This is useful in prioritizing posts at a glance.

Third, you can use discrete numbers or codes in the thread name so your case management system can be linked in. Some systems will automatically generate a case number, so that makes it easy.

Some social engagement systems are integrated with CRM systems that have built-in case management, so it's a good idea to exploit those capabilities so your cases and conversation threads are in lock step.

Conversation threading for customer care is not only possible, but also will become standard for professional engagement platforms. You can take advantage of clustering and other social post attributes to quickly name conversation threads. This makes author searches and outreach much easier for agents and supervisors.

10. Social Care Source Analysis

Social engagement for customer care starts with identifying viable social sources for content. This means organizing content from Twitter, Facebook, LinkedIn, Blogs, and other sources.

Source analysis is the discipline of understanding and adapting to relevancy trends in each source, and doing it well means an uptick in productivity for social care professionals.

What Sources are Meaningful?

Social networking channels are burgeoning with data. But filtering that data into meaningful and actionable information is a complex discipline. Many social engagement tools have been developed to transform this raw data into actionable information. Take, for example, Natural Language Processing / Understanding (NLP/NLU) engines, and keyword search engines.

Depending on the charter of your customer care group, you may focus more on tweets than blogs. Blogs offer rich data and detailed descriptions of news items, trends and opinion. But blogs do not always contain actionable information from a customer service view. Tweets on the other hand, are by nature very succinct, and therefore get right to the point when it comes to a cry for help or a complaint about your product or service.

You can also argue that tweets can carry a lot of noise. Especially re-tweets. Re-tweets are kind of like echoes in the forest after an "original bird" makes a statement. In terms of action, it makes no sense to drill down into a re-tweet because it is usually only the sentiment of the original author that you can take action on.

Clearly, it is interesting to observe trends in re-tweets if you are doing brand and trend analysis, but for a customer care agent, re-tweets are mostly spam.

Making sense out of Sources

First, get familiar with your social analytics tools. Great social analytics tools allow you to not only choose intervals, but also source, sentiment, and outreach progress. If you don't have a good analytics package, consider using an all-in-one social engagement platform that rolls analytics, NLP, and outreach together.

Second, use automated clustering to characterize the persistent business issues and trending topics associated with each source. For example, your analysis may show that 80% of the blogs your search criteria is tuned-in to are news-related or contain product review information.

Further, your cluster analysis may show that tweets on the other hand are mostly service or complaint-related. There are no hard and fast rules. Each enterprise has its own way of profiling sources, so you will have to establish a methodology for profiling based on initial and ongoing analysis of clusters.

Third, focus on sentiment. It's a good idea to see which sources are running "hot" or "cold" sentiment-wise. You can characterize sentiment easily by putting posts in three bins: Happy, Neutral, and Not Happy. Use your analytics package to display each source and the breakdown of Happy, Neutral and Not Happy posts. This will give you insights into prioritization. For example, if you discover that tweets in a certain cluster are running "hot" but the same cluster in blogs is "cold" or neutral, you can use your rules engine to prioritize tweets for "first outreach." This is a great way to mobilize your social care team around the most important authors.

Lastly, you can prioritize and filter sources based on influence. A common approach here is to use your analytics package to list the top authors by the number of posts and their influence. But be careful, just because a tweet is authored by someone influential does not make it a candidate for action.

Influence is an attribute that helps you to anticipate the way an author may respond to a general announcement or even a generic outreach meant for a broader audience.

It's a good idea to follow influencers and try to have direct message conversations with them.

This establishes trust and in turn, any open communications you may post on a certain subject are less likely to be put down in public by that author. Nothing beats sincerity though, so you must be sure to personalize and carefully craft responses to influencers because your brand's credibility is on the line.

Don't forget the Outliers

There are always exceptions. So the danger in source profiling is the fact that there are outliers in each category that are worth exploring. A business analyst or "social quarterback" can do a great service to the rest of your social engagement team by periodically reviewing outliers and cherry-picking or re-routing posts that are actionable.

For example, let's say you've set up your filters and rules engine to put all neutral sentiment blogs on the "news" cluster in a "to be reviewed" bin. Here, your social quarterback or analyst can use influence and other attributes to eyeball posts for content that may have fallen through the cracks in your first pass filtering.

This quarterbacking function is essential for two reasons:

1) It saves the bulk of your social care team from eyeballing each post and taking away from outreach activities; and 2) your analyst or social quarterback can make course corrections by re-tagging clusters, sentiment and relevancy of outlier posts - thus providing "training data" back to the NLP engine. This practice of re-characterizing some outliers is therefore a great way to save your overall team time and to fine-tune the NLP at the same time. This ensures your automation efforts perfect over time.

You can establish meaningful approaches to Source Analysis for your social engagement for customer care efforts in a few easy steps. By developing a methodology for source profiling, and doing standard analysis on your sources, you can quickly put together strategies for each of your social sources. The use of an analyst or social quarterback role can further enhance your outreach performance.

11. Social Engagement Breadcrumbs

Hansel and Gretel left breadcrumbs behind to find their way out of the forest. Now, modern approaches to social care help professionals lead their clients out of the woods of the social web.

Here are some ideas on how to put these ideas to work for your customers.

What are Social Engagement Breadcrumbs?

Many of us are familiar with the idea of web site breadcrumbs, which are kind of an electronic audit trail of where a customer has been on your web site. Social engagement breadcrumbs follow the same logic.

But the difference is between the two is significant. First, web site breadcrumbs represent data that is controlled by your domain. They represent a trail of landing page visits left by someone who was in your "house."

Similarly, social engagement breadcrumbs are an electronic audit trail, but it's a record made up of data from disparate sources, not just the landing pages on your web site. Social engagement breadcrumbs are aggregated in a database and later rendered for use and display to a social care agent.

For example, hash tags from twitter posts can be used to tag conversations on twitter, and article names or keywords can be used to track blogs.

For breadcrumb tracking to be truly useful in the social realm, you have to have the proper infrastructure to gather up conversations from the same "author" across multiple social sources.

Using Breadcrumbs to Make a Social Pie

First, it is important to recognize that customers who use social media often have multiple social handles or "personas." You can unify these personas in a CRM-like fashion by listing each one in your engagement tool. For example, you'll want to see the customer's twitter handle, blog address and Facebook page.

Second, each customer will have an author identity so each social post, regardless of origin, can be stored. The level of sophistication on social post history storage is diverse. Some platforms offer no structure or search capability so you have to "hunt and peck" for the breadcrumbs.

More modern approaches offer everything from a panoramic visual timeline to a simple searchable list - like an email package does.

Regardless, in order to take advantage of social engagement breadcrumbs, you need a way to store and manipulate the data no matter what the source.

Third comes collaboration. Breadcrumbs are easy if you're working alone in a silo. But if you are part of a social care team, you're going to want to share the notes you took, outreach information, and replies with your peers. How else are you going to keep from stepping on each other's toes?

Some case management systems solve this problem readily, and still others allow flexible conversation threading outside of case management. Either way, the ability to cruise through a particular author's posts and tie-in other care professionals' replies is sublime.

Following the Trail

There are best practices when it comes to following Social Engagement Breadcrumbs. These have been figured out by trial and error with social care pioneers and some are borrowed from old school contact center rules. A few examples:

1. Check the Timeframe. Customers will often post on multiple social networks or blogs about the same topic. If you have the ability to sort on a particular customer's posts by time interval or range, by all means use this capability. Why? Because it is easier to pick up on the nuances of customer ask or complaint if you crosscheck the references.

For example, even though a tweet can include a URL to a longer article, the tweet itself may not have enough information in it to give you the full picture. So also reading a companion post from a blog or Facebook mention can often fill in the blanks. Naturally, having all of those cross-network posts filtered into one bin makes this feasible.

2. Understand the Sentiment Continuum. You can ascertain sentiment manually by reading each post, or you may be fortunate enough to have an engagement platform that automatically ranks sentiment.

Either way, it is a good idea to see if an author's sentiment has changed over time or across disparate social media posts. The benefit of doing this is to "tune in" to the prevailing sentiment, or to be able to understand sentiment trends with the customer.

For example, if a person starts with a rant and after a few replies the rant changes to an intelligent exchange - it's good to understand that before you jump in. Ditto when an even-tempered conversation goes hot.

3. He Said She Said. It's considered a best practice to review what your colleagues have said or committed to the customer already. This is easy if you have a CRM-like capability in your social engagement system because you and your fellow care professionals will be sharing the same database.

For example, you may notice that a colleague outreached to a customer with a direct message tweet. If that direct message was sent from an engagement platform that allows all agents to share a Twitter account (a proxy), then you have *collectively* answered the customer. You and the other agents are collaborating on the same corporate Twitter account. In this respect, you have a unified corporate persona to do outreach to the customer.

Let's suppose that the customer to whom your colleague sent the direct message is still complaining and acting like nothing happened.

In this case, it's possible they did not see the tweet. Or maybe they just want to continue venting. Regardless, you are able to do a follow-up with the customer - providing encouragement that "you" were listening and trying to provide a solution. The idea here is that anyone in your workgroup can follow up with the customer and respond from the same "corporate persona."

Having this collaborative view of the customer is also useful in avoiding problems associated with the old "shopping for a better answer" trick. That's when a customer will contact one agent to get a credit, adjustment, etc. and not like the answer he got and then contact another agent for the same thing.

This is in hopes to get a "better answer." Your ability to take advantage of social engagement breadcrumbs makes it a lot easier to solve this and many other outreach scenarios.

It is a worthy and beneficial practice to bend over and pick up the social engagement breadcrumbs your customers leave behind. Here's where you find big clues about trending topics, sentiment and historical discussions.

Breadcrumbs provide guidance to customer care professionals on how to engage with customers. By taking advantage of these breadcrumbs, you can improve the effectiveness of your engagements and make more customers happy.

12. Improving Agent Retention

Social engagement for customer care is vexed with the same workforce problems experienced in traditional contact center environments. Chief amongst these problems is employee burnout and turnover.

It is not uncommon to have churn rates higher than 30 percent in some customer service centers. I've seen it as bad as 100% or more per year. Social care organizations are no different. In fact, agent burnout can be even more acute owing to social spam and a lack of intelligent filtering in first generation social engagement tools.

The biggest complaint we hear from customer care executives using social engagement products is that 85-90 percent of their agents' time is spent slogging through spam. These numbers have also been disclosed in public conferences from executives at companies like McDonalds, for example. Social care professionals who are in the trenches doing the actual outreach also articulate the same complaint we hear from executives.

In fact, as few as 5% of all social posts on a related care topic end up being actionable or worthy of opening a case or escalating. Now let's take a look at this throughout the lenses of a social care professional.

Colleague: "Sue how is your job in the social command center coming along?"

Social Care Specialist: "Oh, I am really getting burned-out on this stuff. Most of my time is spent reading junk and deleting posts that have nothing to do with my job. I'm supposed to be helping people but most of what I see is spam."

Colleague: "What are you going to do about that?"

Social Care Specialist: "I don't know. Probably quit. This spam stuff is not very fulfilling and I don't think I am getting anywhere career-wise."

The Flip Side of SPAM

In addition to the raw tedium of dealing with social spam (doing manual review and deletion), this waste of time chews away at the time that should be spent solving customers' problems. In essence, only 5 - 10% of a social care agent's time is spent doing legitimate outreach to customers.

It's no wonder it is hard to show an ROI on social care initiatives. In effect, put in old school terms, the "agent occupancy" is below 5%.

Enter Natural Language Processing

Most social care executives are well aware of the time-wasting problem of social spam. But many don't know there is an alternative and a solid solution to the problem. The solution is to leverage Natural Language

Processing (NLP/NLU) technology. In some cases, you can bolt it on to a first-generation social engagement system. Another choice is to swap out your system with one that has built-in NLP capability.

An NLP Engine can be "taught" what is topically relevant and what is spam. In turn, the NLP engine can automatically tag certain social posts as spam. With sophisticated filtering and dispositioning software, you can even put social spam in folders like you can do with email so agents don't have to read it all.

In addition to spam, social care agents are also vexed with a lack of prioritization tools. Traditionally, each post, even if it is not spam, has to be manually read and characterized by the agent as being not important, important or urgent. At least if you get rid of the spam, you are eyeballing posts with a greater hope of actually helping someone.

Now imagine if you could use a rules-based engine to automatically tag posts with priorities based on a list of criteria that is important for your business. For example, tagging priorities based on sentiment, influence, cluster, or age of post.

This means you could make all angry customers who are complaining about a new product launch top priority. And this can be done automatically.

Hunt and Peck

Social care agents are also bogged-down with manually hunting and pecking for "like mentions" from the same customer. This can be solved by using a CRM-integrated system, or by using a stand-alone engagement system with CRM-like qualities. For example, the ability to list all posts - regardless of source - from one customer makes it a snap to follow what that person has being saying over time. Even better, if you can sort by conversation thread, social care agents can "get right down to business" and pluck out the most relevant posts in order to figure out what to do.

What's an Agent to Do

Speaking of what to do, and borrowing from the time-honored best practices of traditional contact centers, agents love to have pre-written KB articles, scripts, and next best action suggestions.

Very few social customer care systems offer these natively, but you should consider this seriously for your social care team.

Imagine fewer conflicts with your legal department because you have a basis for "conforming" answers. Imagine consistent instructions based on your business rules. Imagine getting rid of the mayhem and embracing best practices and smoother workflow.

You can stem the tide of agent turnover and burnout by taking simple steps to outfit your social care organization with the proper tools. The near elimination of spam, post tagging and prioritization, and agent assistance solutions are all available and surprisingly affordable. Take a good look at what's available and put a stop to the exodus from your newly formed social care team.

13. SLAs and Essential Analytics

Having effective analytics at your fingertips is essential when it comes to motivating, coaching, and improving your team's effectiveness.

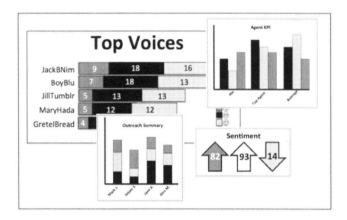

This chapter concentrates on the essential metrics you will need to focus on in order to get a handle on performance and goal setting.

SLA Basics

Before we dive right in to the different types of
reports you should be building, let's first get down to SLA
basics and review some of the common terms we will be
using in this chapter.

Start Time

We can start from the very beginning of when a
particular social post item enters into your engagement
system. This is called the "Start Time." This is the
timestamp for when a social post item first entered into the
social engagement platform and before the item is
assigned to an agent. (See the figure below called SLA
Timeline Anatomy).

It is important to understand that there are built-in
delays between the actual publish date (when an author
actually posts an item) and the time that post enters into
your engagement system. This is owing to the delays
introduced by aggregators or other social feed providers.

These delays can be a few minutes or sometimes hours depending on network traffic.

The easiest parallel to draw here is with a telephone-based contact center. The "start time" is when the ACD answers the call, not when the person started dialing the phone. I mention this because you don't want to penalize agents or groups of agents based on their overall performance if you are looking at publish date. The publish date is important to know, but it is not always representative of when agents actually had the opportunity to engage.

Figure: SLA Timeline Anatomy

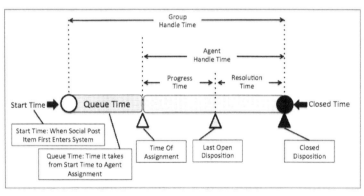

Queue Time

Once the social post item has been logged by your social engagement platform (the start time), the next part of the SLA timeline is the Queue Time. This is a Group SLA metric. Queue time is the amount of time between when a social post item first entered into the social engagement platform (start time) and when the social post item is assigned or selected by an agent (the time of assignment).

Depending on how your engagement system is set up, the incoming queue of social post items will be treated differently. Although the nomenclature may change from vendor to vendor, here are the top three routing models:

• Universal Queue Engine (ACD Model)
• Supervisory Filtering (Controlled Agent Filters)
• Agent "Cherry Picking" (Agent Autonomy Model)

Universal Queue Engine (ACD Model)

With a universal queue or ACD (Automatic Call Distributor) model, social post items are routed based on a variety of attributes. These attributes are both agent and queue-based but also include attributes about the social posts themselves.

On the agent side, these attributes may include the agent's availability, the number of outreached items an agent may have already received in the past hour, and the agent's skill. Even the level of skill proficiency can be taken into account with some systems. On the social post side, information like the publication date, author, and content are used as routing criteria.

With this ACD model, the agent does not decide what social post item to work on. Instead, the universal queue engine (aka ACD) makes the decisions and each item is then pushed to the agents one at a time.

For this model to work, the social engagement platform must provide real time information the universal queue platform. For example, the social engagement platform will have to tell the ACD all about the social posts that are coming in from the social web. For example, the social engagement platform may provide spam scores for each item, sentiment, author influence, and even cluster information. Of course, none of this is possible if the social engagement platform does not have an NLP (Natural Language Processing) engine to do advanced filtering.

The universal queue engine uses its own algorithms and pre-set rules to determine which agent should get a particular post. For example, the ACD could be programmed to send all angry customers who are posting about a particular product to the "retention" queue. In some cases, the ACD could route to a specific individual who is the only subject matter expert on that product.

The level of sophistication in your routing tables is really up to you as a user or practitioner. Many ACDs are set up to simply route based on availability or who has taken the least number of transactions. All of this is programmable.

The are several benefits to using a universal queue model for routing. First, you can "blend" media types. This means the ACD can send a phone call to an agent, and then after the call is done, it can send a social post item, or even an email. This is beneficial because the non-real time items can be distributed in such a way that the "peaks and valleys" are taken out of the equation. That is to say that emails and social items can be pushed to agents who have inbound telephone call downtime.

The second benefit of the universal queue model is the ability to piggyback on supervisory coaching and oversight that already exists in the enterprise.

If your social engagement team is growing, this is important because a ratio of 20:1 agents to supervisors is a best practice. It makes sense to cross-train both agents and supervisors if you are going with a universal queue model for economy of scale.

Supervisory Filtering

Although not a full-blown ACD, supervisory filtering provides you with a fairly sophisticated routing approach. In this model, supervisors are able to change the filters that govern what is displayed on each agent's screen. This means supervisors can be selective about the types of social posts each agent works on.

For example, the supervisor can set an agent's filter so he sees only angry customers who's social posts are on a certain subject. Likewise filters can be set to show a certain workflow disposition or relevancy score. Let's say you are running a grocery store chain.

You could make all social post items about rotten fruit with angry customers who were referred to an escalation group. This is not a full-blown ACD model, but it allows supervisors to create special filters for each agent so he or she has more control over the workgroup.

There are a few benefits of the supervisory filtering approach. First, it's very hands-on and allows the supervisors to quickly adapt to on-the-fly changes.

This is especially useful for smaller groups of agents who are not "blended" agents but rather dedicated to social media. Second, you can deploy this model quickly and there is no systems integration or dependencies on other platforms.

Agent Cherry Picking

Agent "cherry picking" is another model altogether. Here, agents are able to view all of the social post items from a common queue. Agents get to choose which items they will pick out of the queue. This is typically done on a first come, first served basis. In this scenario, as an agent "picks" a post item, it is locked so other agents cannot engage with that customer.

If your social engagement for customer care platform does not do native spam filtering, clustering, author sentiment, or author influence, the type of routing model almost becomes moot.

That's because without advanced filtering, it's kind of a free-for-all anyway. However; if you have a system with NLP (Natural Language Processing) that handles sophisticated filtering, agent cherry picking affords you the same level of routing as supervisory filtering. The only difference is you are putting the controls in the hands of the agents.

The benefit to this "agent autonomous" approach is that you are able to give experienced agents some latitude in filtering their own queues.

Agents that are being on-boarded or who are not very experienced will need more supervision. It is possible with some platforms to "mix and match" these routing models. For example, you can use supervisory filtering for new and inexperienced agents and cherry picking for the veterans. No matter what routing model you choose for your customer care initiative, consider the flexibility and benefit of a hybrid model that allows you to make changes on the fly.

If you have a traditional contact center operation with blended agents, using a universal queue model is a great idea so they can all work on social post items. If you have a group of care specialists who are dedicated to social, you can use supervisory filtering off to the side of an ACD.

During transitions from dedicated to blended agents, you may need all three models, so consider platforms that allow you to operate all three.

Progress Time

Progress Time is an agent KPI metric defined by the period of time between the time of assignment of a social post item and the time set for the most recent (last) open disposition.

For example, you could be assigned an item at noon. It may have taken you until 12:15 to give the item a disposition "doing research," and until 12:30 to give it a disposition of "outreach in progress." You may set up your system such that a disposition of "outreach in progress" is the last open disposition before you close the item as "outreach completed" or "resolved."

This is an important KPI metric because it indicates how long it takes to assess a post item and to begin actual productive work on it.

Resolution Time

Resolution Time is an agent KPI metric defined by the period of time between the last open disposition and the time of final resolution or closed time. The closed time is time-stamped by a closed disposition.

Handle Time (Agent)

Agent Handle Time is the period of time between the time of assignment of a social post item and its being set to a closed (resolved) disposition.

Another way of looking at this is the agent handle time is equal to the progress time plus the resolution time. This can be expressed as an average for an agent based on multiple social post items, or as an absolute number for an individual social post item.

Handle Time (Group)

Group Handle Time is the period of time between the time of assignment of social post items and all of them being set to a closed (resolved) disposition.

This is typically expressed as an average for an entire group of agents. Another way of looking at this is the group handle time is the sum of queue time plus progress time plus resolution time. Alternately, group handle time is equal to the agents' handle time plus the queue time.

Reports vs. Dashboards

There is a distinction between a report and a dashboard. A report (let's say an agent outreach report) provides focus on a specific metric. Reports can often be run based on time frames, intervals, and certain filters. A dashboard, on the other hand, often provides a mash-up of several reports served-up in shorthand. Dashboards often contain several "widgets" – representing a report "snapshot – each having its own theme.

For the purposes of this book, we will concentrate here more on reports. We will identify the most essential reports you will be using. Here is a checklist of the kinds of reports you should consider using. We will cover them here in this chapter:

Table: Report Types

Report Type	Description	Comment
Geographic	Regional or country view of Volume, Sentiment, Profile, Gender, and Cluster metrics.	Useful when engaging on a regional basis.
Cluster Summary	Static Cluster Volume, Sentiment and trends over time	Allows you to focus on the status of persistent business issues
Profile	View of individual profiles by Volume, Sentiment, Region, Disposition, Outreach Status	Good for analyzing outreach by theme or brand
Source Analysis	Tally of original sources of social post items including Native Source, Author Attributes, Sentiment, Volume	Allows for breakdown of traffic and sentiment by their origin.

Table: Report Types (continued)

Report Type	Description	Comment
Volume Trends	View of social post item volume over time across profile, cluster, etc.	Especially useful if you can track volume by sentiment over time
Outreach	Breakdown of outreach activities by agent including volume, spam, sentiment.	Useful to track the "psychographic" of agents by way of their outreach by sentiment
NLP/NLU Analysis (SPAM)	NLP/NLU performance in tagging social post items for Relevancy	Useful in determining how much work the machine is doing with SPAM
Top Voices	Volume, Sentiment and influence of top social posters	Useful if drill-downs to Author history is easy to do

Table: Report Types (continued)

Report Type	Description	Comment
Trending Topics	"Word Cloud" or conversational clusters	Ability to tie clusters to rules-based triggers a must
Workload	Breakdown of status by profile (i.e. in queue, in progress, closed, etc.)	Useful in understanding the SLA time line in real time
Disposition	In-progress and final dispositions and outcomes	Good for a snapshot of where the workload is divided status-wise
Agent KPI	Performance metrics for individual agents	Handle Time, Progress Time, Resolution time all Essential
Group SLA	Performance metrics for groups of agents	Overall Group Handle Time and Max handle time are essential

The Regional Perspective

Geographic analytics help you to focus on metrics on a region-by-region basis. This is possible for several reasons. First, social networking sites can provide the user's city name based on profile information.

This is done by option, but the data can give you a rough idea of where people are tweeting "from." Second, home base for an author can be ascertained based on synchronizing social networking handles with a CRM (Customer Relationship Management) database.

Either way, having this data stored gives you the option of running reports by region. This is important from a brand management perspective, because you can get a quick handle on what trending topics are pervasive in certain regions.

In addition, it is useful to understand how sentiment may be running hot in one region and cool in another.

If sentiment is running hot in one region, but not another, this could be anything from people mimicking each other's sentiment (empathy) to you having a customer service problem in one of your regional care centers.

That is to say if you have your service engagement divided by region, and if the personnel in one region are making people happy, but not so the other one – regional report can reveal this anomaly.

Cluster Analytics

Cluster analytics help you to focus on the status of persistent business issues. I draw the distinction here between persistent business issues (Static Clusters) versus trending topics (Conversational Clusters).

Examples of Static Clusters are "Billing," "Service," "Rant," and "Sales." These are all business-oriented issues that never go away. They are persistent. This is in contrast to trending topics that change all the time. Take, for example: "Can't wait for grapefruit to be in season;" or: "Star fruit is in short supply this week."

Cluster analytics can be very useful in mobilizing around certain business issues or event conversation topics. Let's say for example a cluster report reveals that the "Service" static cluster consistently has twice as many social post items as does the "Billing" cluster.

It would stand to reason, then, that you'd want to have fewer billing specialist versus service specialists. Either that or you've chosen more of a generalist approach to business issues. In this case the issue becomes one of emphasis and priority – that is which post items get worked on first.

Cluster reports are also useful when they can show both the historical totals versus the current period total. It's also a good idea to filter on post relevance (spam score) too. This lets you see which clusters need the most coaching from a social quarterback. Some cluster reports can even be broken down by spam percentage.

An important metric to also consider when viewing cluster data is sentiment. Some cluster characterizations will become obvious right away. For example, it is natural for "Rant" and "Service Problem" clusters to run "hot" – that is have more negative sentiment than other clusters.

Keep an eye on significant fluctuations in sentiment across clusters. If a primarily neutral cluster starts running hot, it's time to dig in and figure out why. You should be able to filter by cluster in real time using your engagement console.

One more tip in the cluster analytics area... Consider the review of trending conversations in addition to static clusters.

If you notice a certain conversational cluster gaining momentum with a lot of volume, you may want to consider using business rules to automatically disposition certain posts that fall under that conversational cluster.

For example, you may get 100 posts a day dealing with a fruit grower's strike in California. These may be legitimately "caught" by keywords in one of your profiles, but nonetheless not actionable by your social agents.

If this is the case, it might be worth writing a simple rule to automatically close such posts as "not applicable." This could save your staff untold hours is lost productivity, so as you can see it is important to pore through your available analytics so you can be on the lookout for labor-saving actions.

Profile Analytics

Profile analytics allow you to view statistics on individual profiles. This can be done by volume, sentiment, region, disposition, or outreach status depending on the platform you are using. These are great for understanding outreach by theme or brand. Profile summary reports can provide you with the key attributes of a single profile or all profiles depending on the flexibility of your platform.

Volume information is one of the basics for profile analytics. This is useful, and especially so if you can compare volume of the current period versus previous periods. This gives you a feel for how volume is trending – that is up, down or status quo.

With more sophisticated analytics, you will see icons embedded in the reports showing a green arrow pointed upwards indicating an upward trend from the previous periods and red arrows pointing downward if the reverse is true. These are often augmented with a plus sign or a minus sign indicating the change from the previous periods. In addition, absolute numbers can be planted next to these icons showing the change of volume between periods.

Equally important for profile analytics are sentiment trends. Modern reports will provide a breakdown of volume for each profile color-coded for sentiment. These are often rendered as line charts.

Here, some profile reports will spore arrows depicting the trend for positive, negative, and neutral sentiment. Still others will provide an absolute sentiment number and percentage differences from the previous periods.

Some profile summaries may include gender information. This is available because Twitter, for example, will pass along gender information in an Author's profile. It will either be male, female or unknown. If your system is synchronized with a CRM (Customer Relationship Management) system, you may be able to derive gender from in-house databases as well.

Top sources information is another attribute associated with profile analytics. Here, you will typically see logos signifying the source alongside numbers indicating the number of post items logged from that source during the report period.

Profile analytics will also sometimes give you a view of cluster activity for each profile. This is important to see because the volume and sentiment of clusters as we mentioned previously can give you insights in to managing your workforce and allocating the appropriate resources by profile.

Source Analysis

Source analysis provides you with a tally of original sources of social post items including the native source, author attributes, sentiment and volume. While interesting, basic source analysis on the surface does not help you to make a lot of important decisions. Where it gets interesting is breaking down sources into sub-sources. Let's say, for example, you have three Facebook accounts set up for your company. You may have one persona or handle set up to provide customer service. Another Facebook page may be set up to handle general inquiries. Still another Facebook page could be set up to handle a brand-specific interest.

If this is the case, you can actually identify four Facebook sources: One for general posts fitting profile search criteria beyond your own corporate pages (that is the public Facebook fire hose), and three others as described previously. Your understanding of volume and sentiment across these four "sources" can be critical in making decisions on campaigns, agent scripts, trending topics that could be covered in FAQs, etc.

Sure you could manage these four sources with four separate tools, but that gets really tedious. It is much more powerful to manage this from a unified system that consolidates all posts and replies with one interface. Ditto the ability to report on this universally - that is across source and sub-source.

Volume Trends

Volume is an important metric, and it becomes more useful depending on the filters you apply. For example, it may be important to know the volume of a specific Facebook page you manage but not so important to know overall volume from all streams. Consider also using filters to understand the volume of certain clusters (persistent business issues) and profiles over time. These are good bellwethers for staffing "by issue" more accurately.

Another essential filter for volume analytics is sentiment. More advanced volume analytics will let you view color-coded lines for positive, neutral, and negative sentiment.

Some others will also include a trend indicator icon showing if a certain sentiment is trending upwards or downwards. The ability to run a sentiment filter on your volume reports is a powerful way to isolate "hiccups" in service, launches or public chatter over sensitive news.

Outreach Analytics

Outreach analytics are a useful way to understand how agents are dealing with your customers. A solid outreach summary will give you a view of each agent's outreach statistics broken-down by sentiment, post relevancy, and volume.

Sentiment is a particularly useful view of agent outreach because it gives supervisors an understanding of whether or not certain agents are "avoiding" negative sentiment posts, or spending a relatively large effort in dispositioning non-relevant items.

Understanding this data is almost like a psychographic profile of the agent. As a supervisor, you are able to coach agents to work less diligently on non-relevant items and more diligently on making angry customers happier.

Spam Analysis

It is useful to review the performance of the NLP/NLU engine in your platform by looking at a spam analysis report. This type of report will provide you with an understanding of how well trained the NLP/NLU engine is and how it is performing.

A spam analysis report will plot the number of posts marked as "spam" by the machine versus the number of post manually marked as spam or manually corrected as spam.

Naturally, the result you are looking for is that the number of items automatically marked as spam are very high and the number done manually or corrected manually are very low. This means that the machine is doing all the heavy lifting, and your agents are just making course corrections.

With a report such as this you are always looking for trends to ensure that the machine is improving over time at least for the first few months of operation.

If the trend is "upside down" – that is more manual corrections or tags than automatic ones – that indicates a training issue or a technical problem.

Some NLP/NLU engines are set up to "fire" or execute in close to real time. Others are set up to fire every ten minutes or so. If your system uses the later, you may want to coach agents to not manually mark spam on brand new items but instead allow the machine to do that. This will save the agents time and also provide you with better spam scores.

During the first few weeks of system operation, you can expect to see a fair amount manual spam tagging or corrections. That's because it takes time to properly train NLP/NLU engines. Over time, the distance between the automatic spam tags and the manual ones will be greater – indicating the engine is doing a good job of automatically marking spam.

Top Voices

The "Top Voices" of your constituency are the authors (customers) that are the most influential if not the most prolific posters.

Depending on the platform you are using, the metrics used to indicate a top voice may vary; but for the most part, a top voices are authors with a) the most influence; and b) the greatest number of posts.

There are many ways to render a top voices report. The standard view simply lists the top voices in a column with their name and number of posts with an influence indicator next to the name. Another view is to provide a horizontal, color-segmented bar next to the name to show post sentiment breakdown.

An alternate view plots authors on a quadrant chart where the "X" axis represents post volume and the "Y" axis represents influence. For example, if an author has high influence and a lot of posts, he or she could be plotted in the upper, right-hand quadrant. Alternately you can do the same kind of chart plotting volume and sentiment.

A nice feature in some engagement dashboards is the ability to "drill down" into author details by clicking on the chart. Top voices reports can be a useful tool in quickly figuring out who the "thought leaders" are so you can jump to their author profiles and check out what they are saying.

Trending Topics

Trending Topics are sometimes called "Conversation Topics" or "Trending Clusters." These are conversational sub-texts that help you to understand the gist of what popular conversations are all about. Depending on the algorithm being used in the NLP/NLU engine, a trending topic will be automatically identified if the components of the word cloud occur more than ten times.

Of course, more advanced algorithms allow you to tweak this threshold number. The issue here is what constitutes a trend. The ability to change the threshold allows your provider to filter out random noise.

There are several ways these trending topics can be displayed in a report or dashboard. The most popular and intuitive display is a "word cloud" which is a collection of words and phrases arranged with colors and font size to indicate the "strength" of each phrase. A simplistic approach is to simply display these clouds side-by-side or in a stack.

More sophisticated approaches include the ability to manipulate these word clouds so you can bundle them together and name them in a "super cluster." This capability is reserved usually for supervisors, analysts and administrators.

From a social engagement for customer care perspective, the value in eyeing these word clouds is for you to determine which ones are more actionable than others. The ability to tie these trending topics to rules-based triggers is essential if you want to automatically disposition social post items based on word clouds.

For example, if your company is a grocery store chain and your profiles are set up to catch conversations about produce including fruit, you may see some clusters including references to the "Orange Bowl at Sun Life Stadium."

Although the word "Orange" may trigger a "catch" with your key word searches, a word cloud about going to the Orange Bowl would not yield any actionable work for your agents. Therefore, you can use rule triggers to automatically close any items in that word cloud.

Of course there are other was to fine-tune your searches using the Boolean operand "NOT" – but you may not know to do that until you see a trending topic all of the sudden. Depending on the make-up of the trending topic, you can also use rules to "transfer" these items to agents in the marketing department or sales if they are actionable for non-customer service staff.

Workload Summary

Workload summary reports are useful in understanding basic SLA metrics and to get a snapshot about how healthy each profile is.

Modern social engagement for customer care platforms allow you to filter by profile and persistent business clusters, so you can drill-down pretty deep into the essential metrics.

Here's what you can expect in a workload summary:
- Number of social post items still in queue
- Number of items open and in progress
- Items already assigned to agents
- Items that have not yet been reviewed
- Items not reviewed and marked as spam

A useful rendering of this type of report is a stacked vertical bar chart that shows the breakdown of the abovementioned metrics over specific intervals.

This gives you a great heads-up display – especially if color-coded – so you can identify trouble spots immediately.

For example, you would want to see the "not reviewed" metric at a steady level, not "getting fatter" over time. If the number of social post items that are stuck in queue keeps growing there are several explanations for this:

- Your brand is getting more social mentions
- There is a short-term bump in traffic
- Your team needs an efficiency tune-up
- You need more people working as social agents
- You need to tighten-up your profile keywords

Disposition Summary

A disposition summary will give you a snapshot of workflow status. Modern social engagement platforms allow you to create and change as many dispositions as you like. Dispositions were covered in chapter 2, but here is an example of some common ones:

- Open - Researching Item
- Open – Outreach in Progress
- Open - In Wrap-up
- Transferred - Sales Team
- Transferred - Retention Desk
- Transferred - Billing
- Transferred - Tier 2 Support
- Closed – No action required
- Closed – Not applicable
- Closed – Outreach completed

Note that each disposition is preceded by the word "Open," "Transferred," or "Closed." This means there are many possible sub-types for each disposition type. The type called "Open" simply means that a social post item is being viewed by an agent and is no longer in queue. In this open state, the agent could be just eyeballing the post item and deciding what to do. Or the agent could be doing some research on the item before completing outreach.

A disposition summary report will provide you with a number key metrics with which to make important decisions about workflow and staffing.

The most basic reports will render as bar or pie charts that let you view the percentage of each disposition out of all the work items.

For example, you may see that 45% of all work items in the last 24 hours are closed, 20% are work in progress, and 35% are transferred to tier two support.

You may have already established SLAs and KPIs for your groups and agents. For example you may have decided that at no time should less than 50% of work items be closed in any 24-hour period.

If this is the case, the disposition summary report is a useful tool in understanding how your team is performing.

Still more sophisticated reports will allow you to see the number of disposition types over time and the final dispositions. This gives you a handle on where the "fat part of the snake" is in terms of workflow.

Group SLAs

Group SLAs give you a solid handle on the performance of entire agent work groups. Agent work groups can be defined and staffed based on skill or business area. For example, you might have a group of social care agents who concentrate on "technical support," and another group that handles "retention."

There are several ways to view SLAs by group. One is to take a look at the average SLAs by profile or cluster. The metrics for these averages include: a) Average Queue Time, b) Average Progress Time, and c) Average Resolution Time.

Average Queue Time (AQT) represents, on average, the amount of time it takes for agents in a particular group to promote post items from the "All" or available part of the media stream and then subsequently "lock" them or "assign" them. This is based on a "Cherry Picking" or "Pull" model in which the agents are empowered to select their own social post items.

If the routing model you are using is ACD or universal queue routing, the agent is assigned social post items automatically and does not "pick" them. Instead they are "pushed" to the agents.

Either way, the moment the social post items appear in the media stream of agents that is when the clock starts ticking on average queue time.

Average Progress Time (APT) is derived from an average of all post items dispositioned as "open" by an agent group during the period.

In particular, this metric takes into account the time it takes for a post item to move to some "open" disposition from the point of assignment to one of the agents in the group.

This time is averaged with all of the other posts dispositioned as "open" by agents in that Group during the same period. This KPI Metric is important because it shows on average how long it takes for the group as a whole to review newly assigned items and decide what kind of outreach progress needs to be applied.

Average Resolution Time (ART) is derived from an average of all post items dispositioned as "closed" and resolved by the agent group during the period. In particular, this metric takes into account the time it takes for a post item to move from the last "open" disposition to a "closed" disposition.

This time is averaged with all of the other posts dispositioned as "closed" by that agent group as a whole during the same period. This KPI Metric is important because it shows on average how long it takes for an agent group to work issues and get them closed from an "open" state.

Another way to view SLAs by group is by the "worst case" scenario. The metrics for these maximum-based times including: a) Maximum Queue Time, b) Maximum Progress Time, and c) Maximum Resolution Time.

Maximum Queue Time (MQT) represents the longest amount of time it took for the entire group to have promoted a Post Item from the "All" or available part of the Media Stream and then subsequently "lock" or "assign" it. If this metric is unusually high for an extended period, it indicates an overall service level problem that needs to be addressed for that group.

It doesn't matter if the distribution model is ACD-like, or Cherry-Picking or Supervisory Filtering. Either way, the maximum queue time measures how long it took for social post items to be assigned to an agent. In this case, the worst performance during the time period chosen is represented by maximum queue time.

Maximum Progress Time (MPT) represents the longest amount of time it took for that group to have moved an item from the queue to the last "open" disposition state.

This metric is important because it shows the "worst case" from the whole agent group on how long it is taking to review assigned items and decide what kind of outreach progress needs to be applied.

Maximum Resolution Time (MRT) represents the longest amount of time it took for that group to have moved a social post item from the last "open" state to a "closed" disposition. This metric is important because it shows the "worst case" from the whole agent group on how long it is taking to resolve open items.

Agent KPIs

Agent KPIs (Key Performance Indicators) are essential metrics for assessing the efficacy of your agents. In fact, Agent KPIs are also essential to measure your own performance and to compare it to your peers if you are a social customer care specialist. It is important to understand that KPIs are individual metrics, not agent group metrics.

Even though they sound the same as the group SLA metrics, they are different because they pertain to an individual.

These include Average Handle Time (measured differently than by group), Average Progress Time, and Average Resolution Time.

Average handle time is an agent KPI based on the average of all post items dispositioned and resolved by a single agent during the chosen period. In particular, this metric takes into account the "time of assignment" of social post items all the way up until they are dispositioned with a closed status (resolved). This time is averaged with all of the other posts dispositioned as closed by that agent during the same period.

Average Progress Time (APT) is based on an average of all post items dispositioned as "open" by an agent during the period. In particular, this metric takes into account the time between the "time of assignment" of all social post items until they are moved to their last "open" disposition. Times for each post are then tallied and averaged for the chosen period.

Maximum Handle Time (MHT) is based on the single social post item that took the longest amount of time to disposition as closed by a single agent during the chosen period.

In particular, this metric takes into account the time between "time of assignment" and the final closed disposition (resolved). This KPI metric is important because it helps supervisors to decide how agents are responding to SLAs and their workload. Traffic re-distribution decisions can be made based on this metric.

In addition to these popular KPIs there are other more simple ones. These include:

- Number of post items outreached
- Number of post items assigned per agent
- Number of outreached items with angry sentiment
- Number of outreached items marked as spam

As a supervisor, you can compare the number or post items outreached per agent and see the top performers pretty easily. This can help you to decide who needs to be coached a little more or who may need even closer supervision.

The same goes for measuring the sentiment of the social post items your agents are outreaching. If agents are resolving happy and neutral social post items, but not angry ones, that could be a sign of avoidance.

Clearly it is not good for someone on your team to avoid customers who complain. That's why these metrics truly matter as a supervisory tool. It's a good idea to use the sentiment part of outreach reports to stay on top of agent behavior. This will help you to offer the best support. After all, angry customers are typically the ones that need the most help.

GLOSSARY

ACD. Automatic Call Distributor. Also called Queue Management or Universal Queue Management. ACD is a legacy term that refers to the ability for a telephone-based service system to distribute incoming phone calls intelligently amongst customer service agents. The term is also used generically to refer to *any* system that distributes work items or any media type to customer service agents.

Agent. Also called Customer Service Agent or Customer Service Representative (CSR). A person who represents a brand, company, product or service. This includes employees who use social engagement tools to listen to and respond to customers using social media. An agent is part of an overall social care team or workgroup that may have other members such as supervisors, quarterbacks, and analysts.

Agent Assistance Scripts. See Assistance Scripts.

Agent Idle Time. See Idle Time.

Agent Occupancy. This is a workforce metric dealing with the effective work time of an agent. It is typically expressed as a percentage. For example, if an agent's occupancy is 60% that means that for 60% of the workday in which that agent is being paid, the agent is engaged with legitimate outreach and customer interaction. The other 40% can be attributed to breaks, idle time, or time wasted on dispositioning spam.

Agent Wrap-Up. See Wrap-Up

AHT. See Average Handle Time.

API. See Application Programming Interface.

Application Programming Interface (API). A means to provide outside access or control of a particular system. Also, a way programs or servers can talk to one another. These can be proprietary or open. APIs piggyback on existing byte-stream command and reply protocols such as WebSocket, REST, or SOA / Web Services.

In social care, most social network providers offer API access to their data streams and programs so third parties are able to integrate with them. Social engagement software must use APIs from Twitter and Facebook, for example, in order to respond to post items.

APT. See Average Progress Time.

AQT. See Average Queue Time.

ART. See Average Resolution Time.

Assigned Queue. See Personal Queue.

Assistance Scripts. Also called Agent Assistance Scripts. These are short, pre-defined text messages agents use to respond to posts. Agent assistance scripts are often used to pass URL links in order to get around the Twitter character limitation of 140 characters.

Author. In the context of social engagement for customer care, an author is the person who wrote a social post, blog, or article. Authors may have many personas or social handles, but they are still one person even though they may go under different names and handles for each social network.

Average Queue Time (AQT). This can be a group SLA or an individual agent metric. Queue time is the amount of time between when a social post item first entered into the social engagement platform (start time) and when the social post item is assigned or "taken" by an agent (time of assignment). You can apply averages on all of the agents in a particular group, and you can average out individual agent's queue time.

Average Handle Time (AHT). This is an SLA metric representing the average of all post items dispositioned and resolved during a chosen period. In particular, this metric takes into account the start time of a social post item (when it entered the queue), up until it is dispositioned with a closed status or resolved. This time is averaged with all of the other posts dispositioned as closed during the same period. See also Group Handle Time and Agent Handle Time.

Average Progress Time (APT). Average Progress Time is an SLA metric representing an average of all post items dispositioned as "open" by an agent group during a chosen period.

This metric takes into account the time it takes for a post item to move to some "open" disposition from the point in which it was assigned in that agent group.

This time is averaged with all of the other posts dispositioned as "open" by agents in that group during the same period. This KPI metric is important because it shows on average how long it takes for the group as a whole to review assigned items and decide what kind of outreach needs to be applied.

Average Resolution Time (ART). Average Resolution Time is an SLA metric representing an average of all post items dispositioned as "closed" and resolved by the agent group during a chosen period. In particular, this metric takes into account the time it takes for a post item to move from an "open" disposition to a "closed" disposition.

This time is averaged with all of the other posts dispositioned as "closed" by that agent group as a whole during the same period. This KPI metric is important because it shows on average how long it takes for an agent group to work issues and get them closed from an "open" state. See also Resolution Time.

Best Practices. These are policies and procedures established over time based on "what works best" for an enterprise. Best practices are often shared between divisions and industry peers in order to establish a way to make incremental improvements based on what works well with practitioners in a similar situation or industry.

Channel Conversion. A channel is a discrete type of communication supported by a specific media type. For example, email is a channel, social is a channel, and phone calls are a channel. Channel conversion is the art of moving a conversation from one of these channels to another.

Cherry Picking. This is a queue management model term. It refers to the ability for agents to select social post items from a queue at will. This is the opposite of an ACD or push model wherein an automatic routing engine decides what social post items each agent sees. A hybrid model uses supervisory filtering wherein supervisors are able to "lock down" attributes of social post items on an agent-by-agent basis. See also ACD and Supervisory Filtering.

CIM. See Customer Interaction Management System.

Closed Time. This is the timestamp for when a social post item is finally resolved and set to a closed disposition.

Cluster. A cluster is a group of words and phrases that make up a sub-topic of common ideas. In the context of social engagement for customer care, these sub-topics represent the undercurrent of a general conversation shared by many people who are posting socially. See also Static Clusters and Trending Topics.

Conversation Topic. See Trending Topic.

Corporate Influence. This is a measure of how "important" an author (customer) is to the enterprise you work for. From the perspective of a social care specialist, it is useful to understand a customer's "loyalty score," or some other numerical measurement of a customer's importance. This score is gotten via an API or data dip into the enterprise CRM (Customer Relationship Management) system. See also Public Influence.

CRM. See Customer Relationship Management System.

Customer Interaction Management System (CIM).
A CIM system provides the infrastructure and support for a customer care team to respond to customer interactions, regardless of media type. Some customer relationship management systems also qualify as CIM systems if they provide support for interactions across multiple channels.

Customer Relationship Management System (CRM). A CRM system provides the ability to access customer records and keep track of what agent said or did what for whom. This may include case management, support for different media such as email and chat, and knowledge base access. CRM systems have become more robust over the years and are just beginning to integrate with social media in an effective way. Some stand-alone social engagement for customer care systems have built-in CRM attributes. Examples of vendors in the CRM space are Oracle, Salesforce.com and Microsoft.

Disposition. A status indication or trigger of a workflow milestone. For example, there are different states a social post item can be in as it relates to customer care.

A social post item can be in an "open" state, a "transfer" (or handoff) state, or a "closed" state. Each state can have many different conditions. For example "item being researched," or "outreach in progress," or "pending approval" are all open dispositions. Likewise, "not applicable," and "resolved," are closed or final dispositions. An Agent will disposition social post items in order to move the workflow forward for an author (customer).

Engagement. This is the act of outreach or interaction with your social constituency. In the context of social engagement for customer care, engagement is the response of an enterprise agent to a social post. Engagement may come in the form of a reply tweet, Facebook post, Blog comment, or direct message. In some cases, especially if already condoned by the customer, engagement may come in the form of another media channel such as a phone call or email.

Engagement Pane. Also Engagement Window. See Response Pane.

First In, First Out (FIFO). First In, First Out is simply an ordering model in which the newest items are listed first.

In the context of a social engagement arrangement, an agent may be able to view post items in a media stream based on this ordering.

FIFO. See First In, First Out.

Handle Time (Agent). This is the period of time between the time of assignment of a social post item and its being set to a closed (resolved) disposition. Another way of looking at this is the agent handle time is equal to the progress time plus the resolution time. This can be expressed as an average for an agent based on multiple social post items, or as an absolute number for an individual social post item. See also Handle Time (Group).

Handle Time (Group). This is the period of time, typically expressed as an average for an entire group of agents, between the time of assignment of social post items and all of them being set to a closed (resolved) disposition. Another way of looking at this is the group handle time is equal to queue time plus the progress time plus the resolution time. Alternately, group handle time is equal to the agents' handle time plus the queue time. See also Handle Time (Agent).

Idle Time. Also called Agent Idle Time. This is a presence and workflow metric indicating the amount of time a customer service agent is not doing any legitimate work, but rather waiting for work to be assigned. In some automated routing systems (aka ACDs), idle time is automatically placed between each transaction so the agent has some "down time" between working with customers. See also Wrap-Up.

Influence. See Public Influence and Corporate Influence.

Key Performance Indicator (KPI). A Key Performance Indicator is a metric used to measure the effectiveness or productivity of an agent in a customer care scenario. For example, the number of customers dealt with or "outreached" during a work period is a standard KPI.

Klout Score. This is a measure of public influence based on the company called Klout, which measures social influence by counting authors' tweets, followers, Facebook friends, LinkedIn connections, etc.

KPI. See Key Performance Indicator.

Last In, First Out (LIFO). Last In, First Out is simply an ordering model in which the oldest items are listed first. In the context of a social engagement for customer care arrangement, an agent may be able to view post items based on this ordering.

LIFO. See Last In, First Out.

Lock-Down Filtering. See Supervisory Filtering.

MAT. See Maximum Assign Time.

Maximum Queue Time (MQT). Maximum Queue Time is an SLA metric representing the longest amount of time it took for the "worst served" social post item to be stuck in the queue before it was assigned or cherry picked by an agent in that group. If this metric is unusually high for an extended period, it indicates an overall service level problem that needs to be addressed for that group.

Maximum Handle Time (MHT). Maximum Handle Time is an SLA metric representing the single post item that took the longest amount of time to disposition and resolve by a single agent during the period.

This takes into account the time it took for that post item to appear in the queue up until it was dispositioned with a closed status and resolved. This KPI metric is important because it helps the supervisor to decide how agents are responding to SLAs and workload. Traffic re-distribution decisions can be made based on this metric.

Maximum Progress Time (MPT). Maximum Progress Time is an SLA metric representing the longest amount of time it took for an agent group to have moved a post item from the queue to its last "open" state. This metric is important because it shows the "worst case" from the whole agent group on how long it is taking to review assigned items and decide what kind of outreach progress needs to be applied.

Maximum Resolution Time (MRT). Maximum Resolution Time is an SLA metric representing the longest amount of time it took for an agent group to have moved an "open" social post item to a "closed" disposition. This metric is important because it shows the "worst case" from the whole agent group on how long it is taking to resolve open items.

Media Stream. A media stream in the context of social engagement is the view a social care agent has of social posts. Depending on the type of engagement platform the agent is using, the media stream may look like an email inbox or it may look like a colorful Twitter "tweet" stream.

MHT. See Maximum Handle Time.

MPT. See Maximum Progress Time.

MRT. See Maximum Resolution Time.

NBA. See Next Best Action.

Natural Language Processing (NLP). Also called Natural Language Understanding. This is a programming discipline under the broader category of artificial intelligence that deals with the understanding of words, phrases, and language. Natural language processing is used in speech-enabled interactive voice response systems in which the automated service responds to spoken commands of the user.

In the context of social engagement for customer care, NLP is also used to automatically determine the sentiment of an author, and it is used to define clusters and topical relevancy scores.

Natural Language Understanding. See Natural Language Processing.

Next Best Action (NBA). An agent assistance tool. NBA are agent-facing scripts that suggest how an agent is to take action for a customer. Next best actions are triggered based on rules that are pre-programmed into a rules engine. They are based on enterprise policies and workflow rules.

NLP. See Natural Language Processing.

NLU. See Natural Language Understanding.

Outreach. A customer care term indicating a deliberate action by a care professional to engage with a customer. In the context of social engagement for customer care, your social care agents are engaging with the authors of social posts.

Outreach activities include answering a social post item in order to address a customer issue, or providing follow-up services to resolve an issue.

Persona. The outward identity established for each social networking instance. For example, an enterprise may create a Facebook fan page for support and another one for sales. Each of these has an outward attitude or feel that is different from the other. The same can be said for personal Twitter accounts or personal Facebook accounts. As an individual, you may establish a business-oriented content on one and a friends and family-oriented brand of content on the other. These different identities are called personas.

Personal Queue. Also Called Workspace or Assigned Queue. A personal queue is a virtual workspace representing the list of work items available to you or assigned to you automatically. In the context of social engagement for customer care, a work item is a social post item you will read, and respond to or otherwise disposition in some way.

Post Item. Also Called Social Post Item. A post item is a unique social post attributed to a single author.

From a customer care standpoint, a post item is also a work item – that is a placeholder representing some action an agent has to take. That action may be to simply read and take notes on the item, or respond to the author, or re-direct the item to someone else.

Post Tagging. See Tagging.

Priority Tagging. See Tagging.

Profile. A profile represents a brand, or a concept, or a product that you define when setting up or making changes to your social engagement for customer care system. Profiles are defined by key word searches you set up. The key words and phrases that you set up as your "social search" criteria act as a dial on a radio, so you can tune in to social posts dealing with a certain subject or brand.

Boolean operands such as "OR" and "AND" and "NOT" are used to create search formulas that make-up your profile. For example, the phrases "ABC Groceries" AND "Fruit AND Vegetables" will be used to fetch social post items that include those phrases.

Progress Time. This is an agent KPI metric defined by the period of time between the time of assignment of a social post item and the time set for the most recent (last) open disposition.

Provisioning. In the context of social engagement for customer care, provisioning is the act of service creation, or setting up your social engagement system. This first requires the collection of lots of data including agent roles, skills, disposition definitions, profile definitions, etc.

Public Influence. This is a score which roughly estimates the influence a social persona has on other people. The most popular is the "Klout" score. These are compiled by looking at the number or posts, tweets, follows, pins, tips, etc. people have on a variety of social networks. See also Corporate Influence and Klout Score.

Quarterback. A social quarterback is a member of an overall social care team or workgroup that may have other members such and agents, and specialists, and analysts.

The quarterback's role is to provide coaching to other team members by recognizing trends, anomalies and hot topics that require the team's mobilization. See also Supervisor.

Queue Time. This is a group SLA metric. Queue time is the amount of time between when a social post item first entered into the social engagement platform (start time) and when the social post item is assigned or "taken" by an agent (time of assignment).

Resolution Time. This is an agent KPI metric defined by the period of time between the last open disposition and the time of final resolution or closed time. The closed time is time stamped by a closed disposition.

Response Pane. Also Engagement Pane. This is an area of the agent's user interface on a social engagement platform where responses are typed or automatically populated and sent back to the author of the original social post.

Rules Engine. Decision-making software that operates based on triggers and rules established by the enterprise.

For social care, rules engines are used to filter, route and prioritize post items so they can more easily be dealt with by the social care team.

Sentiment. In the context of social engagement for customer care, sentiment is a measurement of an author's emotional state or mood. For example, an author may project a happy mood, a sad mood, or a neutral mood. Sentiment is a key metric in a customer care scenario, because this attribute can be used to effectively route or tag post items for agent help.

Service Level Agreement (SLA). Or Service Level Attainment. An SLA is a key metric in determining how effective a care organization is in responding to customer input, complaints, requests and general outreach. SLAs are used by supervisors to understand what part of the operation is doing well and what parts need improvement. For example, average handle time is a key metric.

The assignment of an internal service level goal can be translated into an agreement. For example, there may be a commercial agreement between certain customers and an enterprise regarding responsiveness.

Skills-Based Routing. This is the discipline of automatically routing work items to agents based on their subject matter expertise or skill. Some software systems also allow you to route based on the level of proficiency of a skill, not simply the possession of a skill. Skills-based routing is also used in conjunction with availability routing – that is routing to the next available agent. See also ACD and Universal Queue Management.

SLA. See Service Level Agreement.

Social Post Item. See Post Item.

Social Persona. See Persona.

Static Cluster. There are clusters or "like issues" representing persistent business issues. These persistent issues are opposite of "trending clusters" because they are word clouds dealing with everyday items. For example, a cluster called "Service Problem" is a generic but persistent business issue. Likewise, clusters called "Billing," or "Technical Support" are examples of static clusters. See also Cluster and Trending Cluster.

Start Time. This is the timestamp for when a social post item first entered into the social engagement platform and before the item is assigned to an agent.

Supervisor. A Social Supervisor is a member of an overall social care team or workgroup that may have other members such and agents, and specialists, and analysts. Like the quarterback, The supervisor's role is to provide coaching to other team members by recognizing trends, anomalies and hot topics that require the team's mobilization. See also Quarterback.

Supervisory Filtering. Also called Lock-Down Filtering. Supervisory filtering is an advanced control mechanism for social supervisors to govern the types of social post items each agent sees. For example, a supervisor could choose to set filters for Frank, Joe and Sally to only get angry, at-risk customers who are complaining about a specific product. On the other hand, the supervisor could set filters so Bill, Amy and Chad only get social post items from highly influential authors (customers) who are blogging about a certain brand. Although not a full-blown ACD or universal queue management system, supervisory filtering is a very effective means for doing a blend between cherry-picking and skills-based routing.

Tagging. Also Post Tagging or Priority Tagging. An agent assistance tool. This is the identification of a specific attribute on a social post item. A tag can be done manually or automatically. A typical tag would include a name indicating a specific condition like: "Priority Blogger, Influence Score 50+."

Time of Assignment. This is the timestamp for when either: a) An agent "Cherry Picks" a social post item out of the queue or media stream, or b) A universal queue router (ACD) pushes or assigns a social post item to an agent.

Trending Cluster. (Also called Trending Topic or Conversation Topic). These are "like issues" representing ideas that are not persistent, but rather temporary. A trending cluster represents a current sub-topic of conversation underneath more persistent issues. For example a static cluster called "Technical Support" may have a trending cluster associated with it called "Bad Battery." Trending clusters are often represented visually as a word cloud. See also Cluster and Static Cluster.

Trending Issues. (Also called Trending Topics or Conversation Topic). These represent hot topics that a lot of people are posting about at the same time.

You can filter on many attributes of a trend including geo-specific (where are the voices who are talking most about a trending topic); demographic information (such as authors who share common likes, dislikes), and sentiment (authors ranting or raving about a certain brand). Trending issues are often represented as word clouds. See also Trending Topics and Trending Clusters.

Trending Topic. See Trending Cluster and Trending Issues.

Universal Queue Engine. This refers to software that automatically routes work items based on agent availability, skill and other parameters. A universal queue engine is in essence an ACD (Automatic Call Distributor) that has the ability to route all types of media to agents including phone calls, emails, chats, and social post items. Although ACD is a legacy term, it is still used even though the acronym only takes into account phone calls. See also ACD and Universal Queue Management.

Universal Queue Management. This is the discipline of automatically routing work items regardless of media type. It is closely related to skills-based routing and availability routing. See also Skills-Based Routing.

Universal Resource Locator (URL). This is an address on the internet that represents a service, a web page, or a social network landing page. For example: http://users.twitter.com/Fred123 maybe the URL for a fellow named Fred who has a twitter home page at that address.

URL. See Universal Resource Locator.

Word Cloud. This is a collection of words and phrases presented in a graphical way. Word clouds represent conversations that are happening on the social web. In the context of social engagement for customer care, word clouds are rendered to show trending topics (conversation topics). They are also referred to as "clusters." A typical word cloud will indicate emphasis on the most-said phrases by making them more bold or bigger than the other phrases. Colors are also used to distinguish the emphasis between phrases.

181

Word clouds are not always displayed graphically. You can name word clouds and use them as a means to establish "triggers" for rules-based engines. See also: Trending Topic, Conversation Topic, Trending Cluster.

Work Item. This is a placeholder for an agent action. In the world of social engagement for customer care, a work item comes in the form of a social post item. Unless they are "spam" these items are actionable. That is to say the agent will read and take notes on that item, or perhaps discuss it with a colleague or supervisor before taking action, or even transfer it to another agent. See also Post Item.

Workspace. See Personal Queue.

Wrap-Up. Also called Agent Wrap-up. This is a presence and workflow metric expressed in minutes and seconds. Wrap-up is the time an agent spends to do associated cleanup work after a transaction with a customer. This includes any legitimate activities that benefit the customer including back-end fulfillment activities, filling out sales orders, trouble tickets, etc. See also Idle Time.

NOTES

NOTES

NOTES

INDEX